Cold Spring Harbor

...Rediscovering history in streets and shores

Terry Walton

Foreword by Elizabeth L. Watson
Walking Tour by Ellen Fletcher

Whaling Museum Society, Inc.
Cold Spring Harbor, New York

MAIN STREET SKETCHES BY ANNA DAM-VOLKLE

Acknowledgements

This book is a celebration of what is known about Cold Spring Harbor, and an invitation to discover more. It refers gratefully to many a prior publication, it contains gleanings from minds and hearts and archives of participating experts, and it is buoyed by the gifts of many people. ★ Thanks to Liz Watson for her Foreword and cheerful guidance, Ellen Fletcher for her beautiful Chapter 4, Ann Gill, Sam Scott, Willa Davis, and Frank Jackman of the Whaling Museum, Norman Soule of the Fish Hatchery, Steve Arato of the CSH Fire Department, Mitzi Caputo and Robert Hughes of the Huntington Historical Society, Bruce Stillman, John Inglis, Deborah Barnes, and Elizabeth Powers of the CSH Laboratory, Ann Dwyer formerly of the CSH Library, Dave Micklos of the DNA Learning Center, Stanley Klein, Huntington Town Historian, Julie Moffat of the Brooklyn Public Library, Dave Morrison of the LIRR Historical Society, Natalie Naylor of the Long Island Studies Institute of Hofstra University, Mary Jo Hossfeld of the Village of Lloyd Harbor, Peter Sloggatt of the *Long-Islander*, Richard Welch of *Long Island Forum*, CSH merchants Richard Sutton, Damon Friebolin, and William Reller, Robert MacKay of SPLIA, Norman Brouwer of South Street Seaport Museum, Gaynell Stone of the Suffolk County Archeological Association, Alice Bergida and Ted Hilton of West Side School. ★ Special thanks to John Stevenson, Michael Fairchild, Anna Dam-Volkle, and Byrd Platt for their sketches and photographs, to Dan Barbiero, Frances Elder, Mary Ellen Fahs, Phyllis Fritts, Audrey Goldman, Bonnie Kansler, Barbara Towers, Alison Tung, and Jenifer Walton for early readings, and to the book's skilled and joyful designer Inger Gibb.

Every effort has been made toward accuracy of content and art credits.
The Museum welcomes corrections and additional source information.

This monograph has been published through generous grants from the Banbury Fund, from Furthermore, the publication program of The J. M. Kaplan Fund, from the Ann Eden Woodward Foundation, and from Community Sponsors – Mr and Mrs Norris Darrell Jr, Mr and Mrs Thomas P. Losee Jr, Carol R. Noyes, and Dr and Mrs James D. Watson

COVERS: Gaff-rigged sloops and schooners sail in a calm Cold Spring Harbor, as seen looking north from the milldam road - winding past St. John's Church - that was the earlier path of today's Route 25A. The Hewlett-Jones gristmill is at right, with the "Major Jones Beach" sandspit in middle distance. The site of the red cooper's shop (far left) is part of Cold Spring Harbor Laboratory grounds today (watercolor by Edward Lange, c 1880, Whaling Museum Collection).

Contents

Harbor Recollections

The whaleships ordinarily came to anchor in the outer harbor. My father, John H. Jones, built a dock on the east side of the inner harbor to facilitate their outfitting, and I have seen a vessel fitting out at that dock for a three-years voyage to the Arctic; but the great rise and fall of the tide prevented the experiment being a success, and the original anchorage was resumed. The great rise of the tide – some 7 feet – was in one respect an aid outside, for lying at anchor several months, the anchors sank so deep in the mud that the windlasses of the vessels could not start them, and when the chains were hauled taut for the vessel to pull by the rise of the tide, it often took several tides before the windlasses could weigh anchor, necessitating three days in breaking anchor. . . .

At low tide the water largely covered the bottom, and at the deep hole a number of acres were always filled with 5 to 6 feet of water, even at the lowest tide, which permitted a pleasant pastime for young people to fish and secure results worth serving at the table, the incoming tide always bringing in a fresh supply of fish. Occasionally, but at long intervals, one or two porpoises might be seen sporting in the inside water, but as soon as the tide turned to ebb they made for the outer harbor and no effort to stop them ever succeeded, as they dived under or leaped over the string of boats stretched across the narrow entrance to stop their escape.

Looking south to the inner harbor and the village of Cold Spring Harbor, Major Jones Beach jutting out in the distance, Middle Beach sandspit at left. In the foreground, the steam sidewheeler American Eagle *tows a departing whaleship out to the Sound and the oceans beyond, c 1850.*

– W. R. T. Jones, Governor
Wawepex Society, June 11, 1904

Foreword

The tale of Cold Spring Harbor and its legacies is a real "fish story." Of course before you can have fishing – whether it's whale fisheries or fishing for the secrets of life – you must have water, and here the saga began. A harbor deep from north to south, a natural sandspit nearly dividing the shallow inner waters from the broadening outer reaches emptying into Long Island Sound, Cold Spring was carved out by glaciers eons ago. In historic times the Cold Spring River was dammed in three places to power the village's first industries: flour milling and textile manufacture. But it was the Native American inhabitants who had the right name for it: *Wawepex* – "at the good little water place." Water, water everywhere, especially in the form of natural springs bubbling out of the ground. Capped, these became the artesian wells beloved of the local citizenry, from the colonial era until the present. Do-it-yourself bottling of the local waters is a favorite pastime today along both shores of the harbor, and especially down next to the public dock!

Spectacularly rich in natural beauty, its shallow waters ringed by the morainal hills deposited by the last glacier, Cold Spring Harbor also creates the line of demarcation between the present day counties of Nassau to the west and Suffolk to the east, or more specifically, the towns of Oyster Bay and Huntington, both founded in 1653. It is said that Cold Spring Harbor – with its architectural styles so clearly influenced by the settlers come across the Sound from New England – borders the easternmost influence of the New Netherland colonists. Be that as it may, the harbor's two shorelines lie in two different counties but its neighboring residents have always acted as one . . . for the simple reason that the harbor is a central and unifying element, and a single family at one time owned it all. Captain Thomas Jones ("The Pirate," 1665-1713) carried on negotiations with the native inhabitants until he had gained possession of a broad strip of land that stretched clear from the north shore to the south shore of Long Island, straight through what is now the Township of Oyster Bay. His grandson John Jones ("The Miller," 1755-1819) had the good fortune to marry Hannah Hewlett, whose forebears had long been in possession of a tract of land in the Town of Huntington that mirrored that of the Jones family.

So the story unfolding here conveys the excitement that nearness to the harbor of Cold

Spring has always generated – especially in the last century and growing ever stronger as we take stock of treasures, and as the millennium nears. This book's language and art well illuminate the several centuries of Cold Spring Harbor's history. A small measure of the fact that we have indeed begun to note our treasures – architectural and natural both – is the placement, in the last quarter of the twentieth century, of every single road ringing the harbor – Bungtown, Harbor, Main, and Shore – on the prestigious National Register of Historic Places. And in 1997, New York State designated the fifty-mile stretch from Great Neck to Port Jefferson, from Route 25A north to the Sound, as an official Long Island Heritage Area. We proudly celebrate Cold Spring Harbor's prominent place in this historic area of Long Island's North Shore.

– *Elizabeth L. Watson*

Cold Spring Harbor map, prepared by West Side School 7th and 8th graders and published as part of the school's bicentennial celebration.

Introduction

One sunny mid-February day, as visions for this book were newly coalescing, I did what just about everyone else in town was doing too. I went down to see the harbor. It was unprecedentedly warm, and Cold Spring Harbor – both Main Street and water's edge – was exerting its pull again. Clearly, I was not alone in my response.

A white convertible lazed into a parking space in town, top down. Walkers thronged the shops and strolled the street's length, from Post Office to Library Park. At the Shore Road bridge where ducks always gather, people gathered to admire them as if, in doing so, recovering from winter's isolation. And in the shallows just north the tide was so low that the sand spits showed like resolute islands and the snow white egrets were quietly feasting, unaware of lending beauty to the day.

Sitting at water's edge at last, in bright late-morning sun, I found the air so still that the dronings of insects and the cries of gulls were the centerpiece of sound. The harbor was resplendent in light. I was reminded, all over again, of how special a place Cold Spring Harbor is – and of how it must always have been so in the minds and hearts of our many predecessors: the Matinecocks, English and Dutch settlers in the mid-1600s, mill owners and coastal traders in succeeding generations, whaling captains in the 1840s, and by the 1870s the families out from the City on grand steam sidewheelers with fascinating names, to picnic in our groves and dance in our sprawly, elegant shoreside hotels. In the century just ending, our own century, the picnickers were followed in turn by house- and mansion-builders, sophisticated City businessmen, more shopkeepers, more families come to summer or to stay.

Today the legacies from these several centuries are rich beyond measure. The legacies are maintained by history-minded institutions – the Fish Hatchery, the Lab, the Whaling Museum, SPLIA, and the DNA Learning Center, in order of their foundings, and by churches, schools, library, and post office each of which has its own stories to tell. They are maintained too in Main Street's handsome nineteenth-century houses and shops, and by surrounding Route 25A and Shore and Harbor road structures often hidden by change. And they are maintained especially, if imperceptibly, by the harbor itself, with its cycles of seasons and light, and its freshwater springs at hill and shore and center, the cause of it all.

Among our discoveries, in working on this book, is this one: Many secrets still await us. Not secrets at first, actually, but ordinary facts that drifted away with memory over time. Ordinary secrets in family letters and business billheads . . . in the account books of local brickyards . . . in old ship timbers and mahogany cabin panelling once routinely infused into local houses . . . in arrowheads and pottery and ships resting in the mudflats and hidden out in the deeper harbor. Also among our discoveries is that published sources differ on many a fact and name, as folklore slides deliciously toward fact once it is told and told again. We thus reverted to primary sources (split, thanks to the harbor's being the center of things, between the archives of Huntington and Oyster Bay) for hitherto elusive answers.

Among our discoveries . . . is this one: Many secrets still await us. Not secrets at first, actually, but ordinary facts that drifted away with memory over time.

And as a practicality we confirmed the book's scope to be the harbor itself with its shore-edge roads and adjacent Main Street, and its years ending in 1900 just as the Gold Coast mansion era begins. We referred to Cold Spring Harbor as "town" and "village" (it is actually neither of those but a community within the town of Huntington), and we chose Lloyd Neck vs Lloyd's Neck when both are correct.

Among our most wonderful but tantalizing discoveries are the stories, untold here thanks to constraints of space and scope, that were yielded by the ripples of our inquiries. Stories of the wooden ship timbers visible today in the Eagle Dock lagoon at lowest tides: they are the WWI subchaser SC43, sunk on the very site of John Dole's salvage yard . . . or the "three silver bowls" story: the third and uppermost pond above St. John's church was lost in a 1950s hurricane; or, was it inquisitive boys wondering what would happen if they took out just one piece of the dam? . . . or artist Anna Dam recalling treks to the mud flats with her father years ago, boards on their feet, looking for antique bottles and other treasures . . . or local resident Val Jacobsen's story about his bulldozer sinking in the stream flowing through his foundation site in 1977, at 5 Main Street . . . or, more recently, brambly Library Park – tidal flats in the time of the Matinecocks, centuries later filled as parkland – quietly cleared and maintained by members of the Cold Spring Harbor Fire Department. What splendid interconnections in time we have in this place!

– *Terry Walton, Spring 1999*

1 Setting History's Stage - Our Hills and Harbor

Cold Spring Harbor – small village, deep harbor, streets full of visible history and life. It is a place of energy and invitation.

Invitation for sailing . . . walking down Main Street . . . fishing for blues by the side of the road. Invitation for watching a squall race south across the Sound . . . or slowing down for Shore Road ducklings . . . or, on a rainy day, reading old journals and letters . . . delving into archives . . . wondering about times long gone. And thus, in some unwitting and respectful sense, an invitation for responding to the ancient and monumental work of glaciers.

For it was glaciers that carved out our harbor some 18,000 to 20,000 years ago. Glaciers that created its deepwater five-mile north-south length, its high green-wooded sides, its situation so perfect for the glorious risings and settings of the sun. And glaciers that formed the rich Cold Spring Harbor hills and farmlands, gave us abundant springs and wells, and left behind the grand boulders and sand-cliff edges of the North Shore. Thus, walking around in Cold Spring Harbor today, just an hour east of towering Manhattan, we can share a gratefulness for the icy origins of the town.

All of Long Island, of course, has this glacial legacy. Specific examples are Lake Ronkonkoma and Lake Success, which are "kettle lakes" formed by glacier chunks settling in and melting inch by inch . . . Target Rock and Shelter Rock, each a huge New England bedrock boulder swept south by advancing glaciers, and left behind when the ice retreated north again . . . and the teeth of mastodons and woolly mammoths, found in fishermen's nets along the South Shore. And, closer to home, there's the Cold Spring Harbor Whaling Museum's Main Street boulder: dredged from the harbor years ago and the site today, fittingly, of the slate plaque honoring the bold square-rigged whaleships of the Cold Spring Harbor fleet.

Consider the harbor's natural configurations, so excellent for sailors since centuries ago.

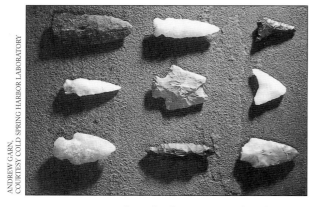

Arrowheads and spear points found on Cold Spring Harbor Laboratory grounds by Lab employee Guy Cozza. Archeological reports confirm artifacts from this site – and others near Spring Street in the village – as being from the Matinecock families who were the first Cold Spring residents.

. . .

This copy of the First Purchase document, 1653, is part of the Huntington Town archives in Town Hall. Thousands of acres from Oyster Bay to today's village of Northport – including Cold Spring Harbor and Huntington – were traded to English settlers by the Matinecock sachem Asharoken.

Beautiful Oyster Bay opens off to the west. Below that lie spacious boat anchorages lending shelter from all but howling northerlies. And, Cold Spring Harbor needs no navigation channel north of its inner harbor. It has plenty of water for sloops racing close in toward the shallows, and its cruising yawls and ketches can head straight to their moorings. The only challenge? Certain rocks and shoals at certain tides, and the sand bar marked by the lighthouse off Lloyd Neck. Nautical charts confirm the assets of this expansive natural harbor.

Our town's water systems, too, are clearly the yield of the glacier's presence. Think of the string of ponds running north along Harbor Road past St. John's Church to the Cold Spring Harbor Fish Hatchery, there paralleled by artesian well-water fresh from its underground sources, and all forever flowing out to the Sound. And, of course, the chilly upwellings that gave the harbor its name; swimmers can attest to these, as their feet encounter the cool mid-harbor surprises.

Then there is Spring Street – just behind Main Street and the northernmost row of shops. Spring Street borders the little spring-fed stream, long ago filled and channeled against the flooding of rain and tide, that is the source of its name. The spring's surroundings were once the site of a flourishing Matinecock village. Early settlers called the modest outflowing Cold Spring Stream.

The St. John's ponds, incidentally, were not always ponds. They once were Cold Spring River – a gathering of freshets running right along today's Route 108 all the way from Woodbury to the Sound. Enterprising residents built dams along it in the 1680s and thereafter, gaining sluiceway power for the early woolen mills, gristmills, and sawmills of a new and growing town.

But the story begins far earlier. . . .

Imagine the look of what would become the island we know, after the thousands upon thousands of years between glaciers and the gradual creation of the Matinecock landscape.

The Matinecock landscape

The first people here were the local Native Americans. They have been on Long Island for at least 10,000 years, notes Suffolk County Archeological Association director Dr. Gaynell Stone, and their descendants are still here. Their ancestors first arrived as nomadic peoples who had walked south from the Bering Strait and Alaska, pursuing the mammoths, mastodons, and rich marine resources, and later the elk herds, of the Ice Age. In Long Island's geographic North Shore region they would come to be known, and well respected, as the Matinecocks.

Imagine the look of what would become the island we know, after the thousands upon thousands of years between glaciers and the gradual creation of the Matinecock landscape. In 1630 and for long after, recounts former Whaling Museum curator Walter Earle in his 1966 publication *Out of the Wilderness*, this land was still one of rich forests and "swamps, marshes and ponds and streams and tidal rivers flowing into deep harbors and shoaler bays; salt meadows along the tidal shores; sweet meadows ripe for grazing, on the uplands and above the beaches," and site of the Matinecock crops of Jerusalem artichokes, squash, wild groundnuts, and sunflowers for seeds and oil. Wolves, bears, deer, and abundant wild fowl lived in the forests and fields. Shellfish "lined the shoal bottoms of the bays and coves – oysters, clams, and scallops. . . . Of equal importance were the springs of cold fresh water, and the fast-running streams – for power, when the time came. It was sweet and pleasant soil."

Matinecock life and the First Purchase, 1653

In 1653, three decades after the Dutch had settled New Amsterdam just thirty-five miles to the west, the Matinecock sachem Asharoken agreed to sell land to a group of enterprising English settlers. This complex arrangement was called the First Purchase; the document recording it survives today as part of the Huntington Town archives in Town Hall. The price for the land, which is today the thousands of acres from Oyster Bay to Northport? An assortment of coats, kettles, hoes, hatchets, shirts, knives, and fathoms of wampum, plus "30 muxes and 30 needles." A muxe is a kind of metal brad or small nail, explains Gaynell Stone, good for drilling holes in shells for stringing as wampum. Our harbor's pearly clam and whelk shells were abundant sources of this prized Native American and colonial currency.

The local Matinecocks were skillful tool makers in quartz and bone. They had hunted game in the lush woods surrounding the harbor, and fished and gathered oysters along the shore, for thousands of years.

The local Matinecocks were skillful tool makers in quartz and bone. They had hunted game in the lush woods surrounding the harbor, and fished and gathered oysters along the shore, for thousands of years. Shell mounds and well-crafted arrowheads and spear points are among evidences of their thriving Spring Street village. The Matinecocks – observed by early colonists as "tall and well built and good to look upon" – were one of Long Island's family groups within the vast Algonquian nation stretching from Canada to the Carolinas. Archeologists describe their arched-pole wigwams as being an inverted bowl shape, made of sapling poles covered in woven mats, bark, or thatching. Their garden crops included corn, beans, squash, pumpkins, melons, and tobacco. And they had chosen the rich North Shore harbors advisedly, for their abundance of birds, fish, eels, oysters, and clams, as well as myriad freshwater streams, and shelter from the sweep of cold northerly winds.

All too little is known of the fullness of the local Matinecocks' lives before European contact. Their existence here – diminishing disastrously from disease and conflict once Long Island settlement began – did not long continue. But the Matinecocks "bequeath[ed] to their small island an abundance of Indian names to be used forever," as local historian Harriet Valentine wrote in 1953. The Connetquot River and the Quakers' Matinecock Meeting House are among these names. Harriet's writings are part of the book *Soundings*, published by the Cold Spring Harbor Village Improvement Society on the 300th anniversary of the First Purchase.

Today, our regard for the harbor must surely be akin to the Matinecocks' own. Their Algonquian word for it was *Wawepex*: "at the good little water place." Cold Spring River was first known as *Nachaquetack*, for Oyster River. Long Island's name, as ethnohistorian William Wallace Tooker writes, was originally *Sewanhacky*, "place of shells." The harbor's seasons and ways are the very center of this community, and must have always been so.

Today too, young explorers still excitedly find arrowheads at the edge of Cold Spring Harbor's Library Park, which is built on the rich river-end marshlands of the inner harbor. Archaeologists' site excavation reports confirm these discoveries virtually all around the harbor's shores and stream edges. Artifacts dating from thousands of years ago on through colonial and far later times – arrowheads and spear points, shell ornaments, pipestems, decorated jar fragments, awls made of bone, quartzite knives, and hide scrapers – are among the documented finds near Spring Street and the Park, and along the harbor's southwestern shore.

Matinecock and colonial artifacts have been unearthed on Cold Spring Laboratory grounds over the decades of its growth. Many of these artifacts have been donated to the Cold Spring Harbor Whaling Museum, for its expanding local history collection. In 1997, the artifacts were part of the Museum's enthusiastically received exhibition entitled "Cold Spring Harbor: Time Measured by the Tide."

Farmlands, mills, and change

The First Purchase of 1653, to which the Cold Spring Harbor and Huntington communities both trace their origins, marked the beginnings of notable European influence. The Cold Spring settlers bought their land and built their houses of hand-sawn planks with thatch roofing. They planted their own fields of corn, beans, squash, and wheat. Interconnections with the

The first West Side School, 1790 – later known as "Bungtown School" for the whaling enterprises nearby – stood beside a big sycamore on the dirt path curving in toward St. John's Church, just above today's Fish Hatchery. It had been moved from its original site just up the hill (today's Route 25A). What is thought to be the same sycamore tree still stands near the church today.

neighboring new settlements at Huntington and Oyster Bay grew apace.

Now came a young town's need for gristmills for flour, and sawmills for lumber (today's Saw Mill Road marks a back road to a 1680s sawmill dam). Before long the mills were joined by tanneries for shoe leather and harnesses, blacksmiths and their forges for nails, tools, horseshoes, door latches and more, and brickyards for chimneys, fireplaces, and houses themselves.

Men dammed the rivers. Mills sprang up all over and some thrived and some did not. Cold Spring River became the ponds that we cherish today, near St. John's Church. It was local carpenter John Adams who built the 1682 sawmill beside the river's upper pond, furnishing lumber for the new generation of houses. Benjamin Hawxhurst built the first woolen mill in 1700 – conveniently siting it at the foot of the lower pond, near the present Fish Hatchery and Route 25A. Bricks came from Jonas Wood's factory, said to have been near Spring Street, 1713, and later from great clay deposits, long operated by the Crossman family, out along Cold Spring Harbor's eastern shore several miles north of town. The harbor's brickworks eventually employed several hundred men, a number amazing to consider for this quiet village today, and from the first their clay formed the jugs, jars, and pottery essential for local households, as well as the bricks that helped to build the town.

Historian Andrus Valentine mentions, again in *Soundings*, some early settlers' strife with the

Matinecocks and even with their Dutch fellow-settlers. There were some frightening times: land disputes, raids, and dark-of-night encounters among them. A century later would come the colonial struggles with England over taxes and the other well-known intolerable acts, all part of what townspeople considered "the pig-headed rule of Great Britain," and finally in 1775 the American Revolution. Scarlet-jacketed Dragoons rode through town in August 1776, heading for Huntington. Local residents were forced to help build Fort Franklin out on Lloyd Neck: prime hilltop vantage for guarding the entrance to the harbor and Oyster Bay. Rumors of British ships were rife. It was remembered as a terrible time, with opposing passions running high.

These several generations, from the First Purchase to the Revolution and beyond, were ones of immeasurable challenge and change. The years brought progress of many kinds to the Cold Spring settlers' lives, including a proliferation of trades, and the first school in 1790 – predecessor to today's West Side School. According to legend, George Washington himself rode by one April morning in his coach with four white horses, just as the school was being built, headed west on his tour of Long Island. Crowds of local men were at work; women and children were raptly watching. Washington stepped down from his coach to greet the townspeople, help raise a rafter for the school, and offer a silver dollar for refreshments all around. "Work went on at the schoolhouse with everyone in fine good humor," note the accounts of the day.

The young Cold Spring was thriving. . . .

Men dammed the rivers. Mills sprang up all over and some thrived and some did not. Cold Spring River became the ponds that we cherish today, near St. John's Church.

Fish Hatchery Hill in Revolutionary Times. Detail, West Side School map.

The Jones brothers' incipient Hicksville & Cold Spring Branch Rail Road (see embossing die) completed grading to Cold Spring Harbor near the St. John's Church pond in 1862, but no rail was ever laid down. . . . The Long Island Rail Road brought service to Syosset in 1854 and to Cold Spring Harbor belatedly in 1868. Meanwhile, it offered perishable produce service to City markets, via Hicksville, Hempstead, and Syosset, as shown in this 1861 advertisement (below).

The Hewlett, Jones & Company 1792 gristmill stood on the harbor's southeastern shore just north of today's Route 25A. Parts of its foundation – with wild shrub roses incidentally present at water's edge – can still be seen today.

Schooners laden with Cold Spring bricks from the Crossman yard, shown here looking west from today's West Neck Road, sailed across the Sound, to the City, and even to Maine – sometimes bringing cordwood for the kilns on the return trip. Brickmaking in the harbor's several yards apparently began before 1700, and employed more than 300 men at the height of the trade. Crossman family records show that seasonal workers received $7–$8 a month, and extra cash for good behavior or for tending the kiln on Sunday. The yard made 11 million bricks in 1871.

2 Whaleships, Mills, and Coastal Traders

What a century was beginning for Cold Spring Harbor! The first mills, rich farmlands sown with care, a harbor of shelter and many new venturings, and the first school in 1790 had set the stage.

The nineteenth century would bring a proliferation of mills for wool, wood, and flour. With them in turn came coopers and blacksmiths in support of bold new business plans, coastwise trading in harbor-built sloops and schooners, and whaling voyages around the world. In this century, lives would be forever changed by new churches and growing schools, and by the railroad and the sidewheel steamer excursion vessels sailing out of the city to our picnic groves and resort hotels of unprecedented elegance. All within just a few generations. All within the still-small Cold Spring Harbor.

The early centuries also sowed the seeds for today, in quiet ways not yet foreseen. Place names are among those legacies. Consider Huntington's Tanyard Lane from the leather tanneries, and the aforementioned Saw Mill Road from the lumber mills; Bungtown Road from barrel-making on the harbor's western shore (bungs were the tapered wooden stoppers for the whale oil and flour barrels of the day); and of course "Bedlam Street" from the multilingual chaos of whaling men swaggering up and down the main street of town.

Consider also Eagle Dock, once the berth of the graceful steam sidewheeler *American Eagle,* as she carried her cargoes to and from New York City right on time. And Spring Street, Oyster Bay, even, best of all, Cold Spring Harbor. Our natural and industrial histories are alive in these names.

Actually, early in the 1800s, we weren't "Cold Spring Harbor" just yet. The place was still "Cold Spring," and became a Port of Delivery by Act of Congress on March 2, 1799 – as an official record-keeping place for sending out and receiving the sloops and schooners of the era's growing trade. But it wasn't until 1826, when there was risk of confusion with the Hudson River's own town called Cold Spring, that the Post Office gave this town its fuller name.

The proliferation of mills

It was the mills, perhaps most of all, that helped the new village grow. The first mills — damming up Cold Spring River for the ponds so innately part of St. John's Church and the Hatchery today — had been mere foreshadowings. Soon had come woolen mills for the fine merino wool of the local sheep, and gristmills for flour, sawmills for lumber. By the early 1800s the town was alive with "mills, mill ponds, dams, and water rights."

One mill, a Hewlett, Jones & Company venture on the harbor's southwestern edge, had a special arrangement that would astound us today. Earlier accounts differ on just which mill it was (Ramona Sammis's *History of Huntington* confirms it as the company's spinning mill on the harbor's west shore, rather than its gristmill on the east). But the mill did receive its water by an ingenious method.

No less noteworthy was Richard Conklin's mill of 1782, which made linen paper of such fine quality it was shipped from Cold Spring to England for use in the fine bibles of the day.

Picture it. The pond water "was brought through a wooden trough, or sluice, across the highway [earlier path of Route 25A's Fish Hatchery Hill] overhead by perhaps 12 feet," notes Walter Earle in *Out of the Wilderness*. "And thence through a canal, or mill race, along and just to the west of the road for perhaps 300 yards, until it came to the mill." Thereafter the water flowed endlessly over the mill's sturdy waterwheels, which turned the mill-stone, which spun the wool into yarn. A sluiceway twelve feet *above* the road. What a clever scheme!

The Hewlett-Jones gristmill jutted out from the harbor's southeastern shore just north of the Jones Corner (today the Route 108/Fish Hatchery Hill) intersection. The mill worked well until long after the Civil War and its trim brownstone underpinnings can still be seen from the water today. Traces of its efficient pond-to-harbor sluiceway, now filled with underbrush, still parallel Harbor Road. No less noteworthy was Richard Conklin's mill of 1782, which made linen paper of such fine quality it was shipped from Cold Spring to England for use in the fine bibles of the day. A blue-and-bronze plaque at the Main Street–Shore Road corner marks its site today, just opposite the Firehouse.

Trades in the growing town

Trades burgeoned in the growing town. Men began the ventures that shaped Cold Spring – among them members of the Bunce, Conklin, Hewlett, Jones, Mahan, Rogers, Seaman, Titus, and Valentine families, and many more. In 1826, as the Whaling Museum's local history exhibits have shown, Main Street took shape as a dirt path full of promise:

"Judge Conklin commenced to sell plots of land along Main Street at the western end of town, while Moses Rogers sold lands along the eastern boundary of the town. Out of these lands grew downtown Cold Spring Harbor, which by 1850 had largely taken its present form with a population of 500 residents."

Local shipyards built the harbor's sloops and schooners. Ships' carpenters repaired them, and sailmakers made their sails. Inland, the farms grew wheat, hay, corn, and other food crops in abundance. Neighboring fields went into pasture land for sheep – which were oftentimes herded straight through town. Blacksmith shops shod the farm and carriage horses, and made latches, hinges, and cooking pots for houses, eel spears and clam rakes for local baymen, and, it is likely, harpoons and lances for the whalers. Coopers made barrels for shipping flour out in the coastal schooner trade, and soon too for whale oil throughout the whaling years here.

The harbor's bricks were trekked out and along the seaboard in local schooners – a thriving coastal trade supplying other growing towns – and formed foundations for local homes. The brickyards would flourish for nearly two hundred years, until there was too little ready firewood for their kilns. Brickworks traces can clearly be seen along Lloyd Harbor shores today – among them a deep pond that was carved over time as a clay pit, the curving shore edge of the inlet where schooners once loaded their cargoes, and the Crossman family house itself. Bricks stamped "Crossman," "West Neck," or "WKH" for William K. Hammond, who later leased the yard from the Crossman family, are still part of many a local brick path or foundation.

The "Jones Family" story begins

It was these enterprises, particularly the mills, the cooperages, and the venturesome whaling fleets that together they supported, that shed early light on a phenomenon of our town: the Jones family.

The story had begun in 1695 when "Major Jones" came to neighboring Oyster Bay.

Walter Restored Jones (at left) and John Hewlett Jones, descendants of the legendary "Major Jones" whose vast north-to-south landholdings once included the South Shore's Jones Beach. Walter R. and John H. were principal founders of the Cold Spring Whaling Company in 1838.

Thomas Jones was an enterprising fellow, a privateer with papers from the British that permitted him to seize ships and their cargoes "for the crown." Major Jones's progeny were outstanding and effective citizens of the town. Two of his great grandsons – John Hewlett Jones (1785-1859) and Walter Restored Jones (1793-1855) – were, with three other brothers and their friends, the energy behind what would become known as "the Jones industries." *Soundings* notes the scope of the Jones brothers' activities:

"Between them, they owned and operated in 1837 a mill that made good woolen fabrics, grist and flour mills, a general store, a barrel factory (cooper's shop), a brickyard, several sloops and a steamboat and repair yards – all at Cold Spring, on or near the harbor." Later research in the National Archives and in Jones family documents suggests that their steamboat *American Eagle* was not acquired, by lease or purchase, until 1842 and that the brickyards were not among the Jones family ventures until 1860. Nevertheless, the lively 1837 endeavors were mere prelude to Cold Spring's whaling trips to distant oceans – newly begun by these enterprising descendants of Privateer Major Jones.

The harbor and the town near mid-century

Envision the harbor and its town in the years just before mid-nineteenth century, as shown so dramatically in the Whaling Museum's c 1850 diorama:

The "inner harbor" – the water south of today's Main Street and the sandspit jutting out from the harbor's western shore – was alive with trade that was timed by the tides. In these years, the harbor marshlands extended well south of today's Route 25A, virtually all the way to the St. John's Pond spillway. The inner harbor may have silted in some since those trading days, but even then, accounts say, the mud flats stranded many a working vessel that missed her tide.

The whaling industry would arise in the 1830s, thanks largely to Jones family investment

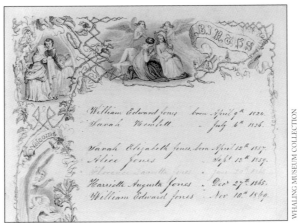

Detail of the Hewlett family bible, recording births starting with that of
William Edward Jones on April 9, 1824. . . . View from the Hewlett house
"Harbor View" on Harbor Road, in an undated watercolor by artist
Edward Lange. . . . The house was built for the first Customs Inspector,
Jacob C. Hewlett, in 1824, and may still be admired by passersby today.

The Baptist Chapel, corner of Main Street and Poplar, was built in 1844-47, and held its baptisms in the harbor in all seasons. The chapel is an attractive private house today, its steeple removed.

and ingenuity. Bungtown Road factories and mills thrived on the harbor's western shore. They milled wool for shirts and blankets of superb reputation. They ground flour for breads, and produced barrels for shipping flour out to New York and East Coast ports, and then whale oil for lamps and soap and more.

The railroad was heading steadily eastward, seeking a straight-through link to Boston by rail and cross-Sound steamer, but was not yet serving Cold Spring Harbor. Coastal trading flourished, and local shipyards were kept busy with building and repair. Beamy sloops and schooners loaded at Wood's dock, which had been built out into the mud flats of the inner harbor specifically to receive them. Farm and dairy goods, hay, wood, bricks, and meats shipped out steadily to coastal towns and west to a growing New York.

Churches and schools responded to townspeople's needs. St. John's Episcopal Church was the century's first for Cold Spring Harbor, built in 1836 by Jones family members and others with essential financial help from Trinity Church in New York. Main Street's Methodist Episcopal Church was founded in 1842, by circuit-riding preachers going town-to-town on horseback and holding services in private homes. In these years, Methodist and Baptist church-goers held glorious afternoon picnics nearby in Titus Grove, a popular place encompassing the present grounds of the Cold Spring Harbor Whaling Museum.

At Main Street's Baptist Chapel, built between 1844 and 1847, families sang the "Sailor's Hymn" on Sundays to keep husbands, sons, and brothers safe on their trading trips in ships and sloops and schooners. Members of this congregation held their baptisms in the harbor regardless of season. In winter, it is said, holes were cut in the ice for this important ceremony. Young schoolboys shouted down from shoreside hills, encouraging additional dunking. The little chapel still stands today, on the southeast corner of Poplar Street and Main, and is a quietly beautiful private house.

St. John's Church, built in 1836. Wagon tracks head past the church in this photograph, c 1880, and across the pond and its milldam road – the original route of Route 25A. Today's Route 108 and Lawrence Hill Road are in the distance.

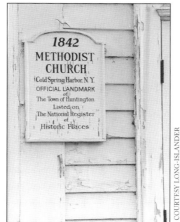

Main Street's Methodist Episcopal Church, founded in 1842 by horseback-riding preachers going from town to town offering services in private homes, is being restored today as headquarters of the Society for the Preservation of Long Island Antiquities.

15

West Side School was forty-six years old in 1836, when St. John's Church was built, and destined to be known as "Bungtown School" because of its proximity to whaling activities on the harbor's western shore. The schoolhouse then stood near the upper road that leads into the church today, just above the Fish Hatchery – the same dirt road then winding past the church and over the milldam and its pond. Schoolboys could gaze out the windows at the busy harbor, carving the ships of their dreams in their cedar desktops: sloops, schooners, whaleships, and more. West Side School still exhibits one of the desktops – its carved fleet still sailing – and proudly remembers its history as "the nursery of sea captains."

Main Street developed steadily in response to need. East Side School was built in 1845, just west of the Main Street–Goose Hill intersection. Eagle Engine Fire Company, a forerunner to Cold Spring Harbor's Main Street Fire Department, purchased its first hand-pulled pumper in 1852. The Post Office at one point operated out of the Jones General Store, later at 16 Main, and then across the street at the corner of Main and Shore, the site of a barbershop

Sheep routinely headed along Main Street en route to pasture.

today. Shore Road would be completed in 1860, bridging the Spring Street tidal pond. The little spring was by now largely channeled or filled, lest it remain a "continual bog" or, in high tides and rainstorms, an unintentional rowing avenue end to end.

News of the contemporary wider world – in the two decades before mid-nineteenth-century – mentioned people and events that are legendary to us now. The news confirmed that different parts of the world were moving at decidedly different paces: Davy Crockett, Napoleon, Queen Victoria . . . writings by Dickens and Hugo, Wordsworth and Emerson, Melville and Marryat and Poe . . . . paintings by Corot, Constable, and Daumier . . . . music by Mendelssohn and Chopin . . . and Darwin's world-changing voyage on the *Beagle* – they all coincided with our boisterous coastal trading and early whaling years.

Out at sea, New York's clippers sped to China for silver and tea, sometimes taking shoes, clothing, and foodstuffs west to California's Gold Rush towns along the way. Packet ships raced transatlantic to Liverpool carrying cotton and grain, and already, in these decades, giving way to the swift steamships that would signal their demise. The Civil War was still to come.

PUBLIC SCHOOL, COLD SPRING HARBOR, L.I.

with love, from Ethel Stoot

CSH FISH HATCHERY

East Side School, built in 1845 on the site of today's DNA Learning Center.

On Long Island, Cold Spring Harbor thrived and other new communities grew right along with us. New York Harbor historian Robert Greenhalgh Albion notes the names that coastal skippers gave the East River's treacherous rocks, as they sailed their cargoes from Sound to City: "Flood Rock," "The Gridiron," and "Bald Headed Billy." The *Long-Islander*, founded in 1838 by poet Walt Whitman, recorded many a detail of Cold Spring Harbor's own venturesome years. The Jones brothers were proceeding with their plans.

Our boisterous whaling era, 1836–1862

Cold Spring Harbor whaling, the industry that would so intensely and long-lastingly shape our town between 1836 and 1862, had been very much the inspiration of John H. and Walter R. Jones. The Matinecocks and early colonists had ample experience in "drifts" (dead whales washed ashore) and "shore whaling" (going out in small boats, harpoons in hand, off Long Island's South Shore and eastern end). But the brothers envisioned lucrative additional possi-

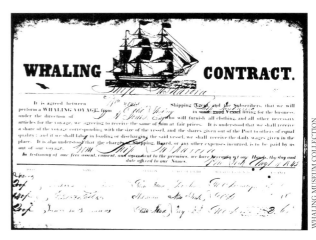

Capt. Eli H. White of Southampton (top left) commanded the whaleships Sheffield, Alice, *and* Tuscarora. . . . *Cold Spring whaleman Capt. Dewitt Barrett; his beautiful Main Street house still stands. . . . Whaling contract for 102-foot full-rigged ship* Tuscarora, *August 1845, specifies payments for fitting out expenses, labor, and cargoes. . . . The 100-foot bark* Monmouth, *first in the Cold Spring fleet, by artist Eric Tufnell.*

bilities: larger markets for their woolens, flour, and ships' supplies. They had considered whaling in the mid-1820s after the waning years of the War of 1812 and the virtual collapse of the textile industry. The first Cold Spring Harbor vessel finally sailed in 1836, and the Cold Spring Whaling Company itself was in operation from 1838 to 1851.

"It must have seemed a likely venture," observed the Village Improvement Society's chroniclers: The brothers "could supply the flannel and heavy woolen cloth to be sold to the seamen aboard ship (at fancy prices), the meal and flour for the galley stores, the barrels, casks, and tubs, and even the bricks for the try-works aboard ship; also sloops for lightering, and the steamboat to transport the oil and whalebone to the city markets – all being ready at hand and subject to little competition." The Jones General Store stood on the harbor's southeastern shore. Mills and barrel factories hummed along on the opposite shore just west across the inner harbor, where the Cold Spring Harbor Laboratory is today. Bungtown was born.

Out of the Wilderness notes, however, that the brothers had not hurried their decision to begin:

"The Jones[es] were deliberate. John H. and Walter R. deliberated for . . . years before deciding to take the plunge. The prospects looked good. There was plenty of water in Cold Spring Harbor for ocean-going vessels to come in and anchor off the west shore; and their neighbors at Sag Harbor . . . Mystic, Stonington, and New London were doing well. . . . Finally, in early 1836, the Jones group decided to go into whaling. They bought, for about $20,000, the small and old bark *Monmouth* [of Boston]. . . , outfitted and stocked her in Cold Spring, and sent her out, in July 1836, on her first cruise as a whaler of Cold Spring. The *Monmouth* was 100 feet overall." She would sail to many a distant ocean, this plucky little bark, seeking the whale.

For the next twenty-six years Cold Spring Harbor actively operated a fleet of nine ships (two barks and seven full-rigged ships): *Monmouth,* followed by *Tuscarora, Huntsville, N. P. Tallmadge, Richmond, Alice, Sheffield, Splendid,* and *Edgar.* The ships sailed transatlantic to the coast of Africa, or down to the South Atlantic and the Indian Ocean. Others braved rounding the Horn for the South Pacific whaling grounds, and some went on to the Pacific Northwest and all the way up to the perilous Arctic. Miles and miles

For the next twenty-six years Cold Spring Harbor actively operated a fleet of nine ships . . . transatlantic to the coast of Africa, or down to the South Atlantic and the Indian Ocean. Others braved rounding the Horn for the South Pacific whaling grounds, and some went . . . all the way up to the perilous Arctic.

Rum and other refreshment welcomed whalers at Van Ausdall's hotel, today the Inn on the Harbor restaurant.

from home, they were, their sailors living lives of doldrums and terrible dangers, freezing in the ice or sweltering southward, losing lives, keeping journals, carving scrimshaw on flat calm days.

When the whaleships finally came home to Cold Spring Harbor, sometimes more than four years after departing, these were grand occasions. Returns were announced by a signal cannon on the harbor's western hillside (site of the Cold Spring Harbor Laboratory today). Townspeople flocked to the shores to welcome the sailors home. And when the bartender at the local tavern heard the signal, legend has it, he plunked shot-glasses of whiskey at the ready all around the bar. Welcome home!

Whaling. It was a brief but heroic era whose legacies shaped our town and touch our lives today. It was in reality a tough and oftentimes exploitive business that brought international commerce into a small town here, and attracted largely non-residents to its grueling shipboard tasks. Maritime historian Frederick P. Schmitt's highly respected book *Mark Well the Whale*, published by the Whaling Museum in 1971, records the harbor's whaling years with engaging style and meticulous accuracy. Other accounts describe them in lively but romanticized terms:

"This was by far the most picturesque and dramatic adventure in the history of the entire community," *Out of the Wilderness* observes: "During the period from 1825 to the outbreak of the Civil War, whaling, for oil and bone, was at its peak. There was no petroleum and no [spring] steel. Whale oil was used for every form of illumination and lubrication, and to make soap and other cosmetics, and many other uses; and whalebone was used for all manner of stiff, resilient articles such as umbrella ribs, buggy whips and stays in women's corsets. Whaling was, actually, one of our [country's] major industries. . . .

"Ashore, there were carpenters making barrels and casks, and repairing spars and broken boats and doing an infinite variety of odd jobs; there were millers making grist and flour; and the workers in the woolen mills; there were sailmakers, making and repairing sails; and black-smiths making all manner of hardware items for the vessels and for the buildings; there were bookkeepers; . . . and finally there were laborers to do the heavy work, such as loading and unloading, digging trenches and foundations, and the like."

Young Helen Rogers' diary, 1843-1850

Cold Spring Harbor in the Jones industries era from 1836-1862, including the time that young resident Helen Rogers records so carefully in her diary of 1843-50, thrived in what has been called its "boomtown years." Helen lived in a beautiful house on the corner of Main and Spring streets, called "the Vineyard" for the grape arbors in its spacious yard. She was a bright-spirited fifteen-year-old when she began writing in 1843, notes *Out of the Wilderness*: "Miss Helen was keen and alert, sprightly and gay, observant and brightly articulate, possessed of a delightful sense of humor." She writes lightheartedly of how she and her young friends spent their days, which surely held more than the carefree hijinks and parties that she records. *Out of the Wilderness* observes:

"Despite the time and the place, the Cold Spring Companies [Helen's name for her group of friends] did pretty well. Helen writes of their going together on hayrides, on sleigh-rides and boatrides – particularly of all-day picnic trips across the Sound to Shippan Point (Stamford)." There are notes of apple-peeling parties, quilting parties, circus parties, camp meetings, amateur theatricals, and dances. And she makes sure to mention the trysting place at the old willow, in what is today's Library Park.

Helen's father Daniel Rogers, an attorney, was deeply involved in the whaling venture. His partners included members of the Jones, Conklin, Underhill, and Willets families: prominent men all. The Rogers diary gives meticulous accounts of ships and the harbor, as noted by historian Harriet Valentine in the Whaling Museum's 1981 book *The Window to the Street*. The book creates the context for the diary, from which it quotes extensively. Of Helen's proximity to Seaman's Dock and Shipyard, Harriet Valentine writes:

"Helen's preoccupation with ships is easily understood, for she could see and hear the busy clamor of voices and day by day the construction of some sleek schooner by Cold Spring men.

"The San Francisco streets were all mud and fleas so thick we can't get no comfort," Valentine *wrote home in 1850. "If I try to get home . . . it will take about 40 days. . . . If I don't see you, try to meet me in heaven."*

Looking east to Cold Spring Harbor, Main Street and Spring Street in distance, c 1850. At right, the shipyards and sail lofts of Harbor Road and the inner harbor, with gaff-rigged sloop and square-rigger conducting the business of the day.

CAP'N
ENOS

*Capt. Manual Enos, whose 1860s Main Street house is known today for its historic
pink shutters, served in the Cold Spring whalers* Huntsville *and* Sheffield. *He came
here from Sag Harbor, later sailed out of New Bedford, and was reportedly lost at sea
with all hands while serving as captain in the whaleship* Matilda Sears *in the 1880s
(Whaling Museum Collection).*

Scores of local men were specialists in shipbuilding: carpenters who formed the great spars, sailmakers, planking gangs, riggers, caulkers, and even caulking boys who ran around with buckets of hot tar, oakum, and other 'Needments.' "

Of particular note in the diary, not surprisingly in light of her father's business interests, is an entry about the *Huntsville's* grand return to her harbor. Surely the *Huntsville's* captain, as was the custom of the day, took his bearings on the bright white steeple of St. John's Church as his ship tacked south from the Sound at the harbor entrance:

"The *Huntsville* arrived April 21 [1849] with 4,200 barrels of oil, 20,000 lbs. of bone to John H. Jones, $40,000 of gold dust for Boston, having in 18 months made the greatest voyage on record." The whale oil and bone were the yield of her voyage to the South Pacific and the Sea of Okhotsk, then on up to San Francisco. Her gold dust came back around the Horn bound eventually for Boston, in the first year of the California Gold Rush. One of her crew members, Manuel Enos, later built a fine house on Main Street in the 1860s; it is still there today, diagonally across from the Whaling Museum, and known respectfully by its pink shutters.

Helen's diary also records the wistful looks in young boys' eyes when the Gold Rush ships came in. Local carpenter Israel Valentine did venture west to the goldfields in 1849, she writes, with several of the most determined of these youths aboard. The trip proved ill-fated: endless rain and mud . . . the schooner *I. B. Gager* delayed five months with essential supplies . . . no gold, and the young boys just running off, all to the great distress of Cold Spring Harbor families. Imagine the sodden-cold chaos of it all: "The San Francisco streets were all mud and fleas so thick we can't get no comfort," Valentine wrote home in 1850. "If I try to get home . . . it will take about 40 days. . . . If I don't see you, try to meet me in heaven." He was back in Cold Spring Harbor, deeply discouraged but safe, by spring.

Back home in these mid-century decades, Harriet Valentine writes, Cold Spring Harbor was most assuredly alive and well: "The hills hummed with noises from the mills, coopering shops, boatyards, and activities of the vessels in the harbor. In fact, at this time in Cold Spring's history, she was larger than her neighboring village, Huntington. The whole community was bent toward supplying the needs of the whaleships, coasters, and their crews. Farmers and baymen also worked from dawn to dusk to supply sustenance for themselves and their robust community."

What grand connections in time are these . . . the 1843 diary, the church steeple, the pink-shuttered house, the West Neck bricks in local paths, the harbor itself, all still here as part of our everyday lives. Helen Rogers' Vineyard house is well cared for today and surrounded by gardens; her diary remains prized by the Whaling Museum.

Two more Jones brothers ventures

It was about this time, in a mid-century era of busy harbor and eastward-heading railroads, that John H. and Walter R. Jones considered still two more intriguing endeavors. One created a legacy that survives in the harbor today. The other – which did not happen – would surely have changed Cold Spring Harbor beyond imagination.

The first new venture was the acquisition of the aforementioned steam sidewheeler *American Eagle*. Steam vessels had begun to challenge sailing ships by the 1830s and would eventually spell the end of their grand and graceful lives. But locally, for now, steam transportation was the answer to the tradesman's needs.

Steamships were efficient transport for goods. Roads were rutted and muddy, and horse-drawn wagons just too slow. Schooners for New York or points east had to answer to weather and the tide, oftentimes disastrously. New York's East River, not a river at all but a treacherous strait whose eddies and "Bald Headed Billy" rocks had wrecked many a stout ship, was a stumbling block to swift commerce. Yet time was, increasingly, of the essence.

The Jones industries answer — steam transport for the town's marketable goods such as hay, bricks, and fresh farm produce. The brothers were operating the *American Eagle* by about 1842; National Archive records confirm that she had been a familiar harbor sight for years before. The brothers dredged mud and sand to build one dock for her near their gristmill on the harbor's southeastern shore (surely a challenge this far south in the shallow inner harbor), then a second one further north. *Eagle* towed sailing vessels against contrary harbor winds whenever needed. And she worked hard carrying freight and passengers in regular steamboat service to New York at Peck Slip, adjacent to today's South Street Seaport Museum.

The sturdy 122-foot steamship American Eagle *towed Cold Spring whaleships harbor-to-Sound and back, in contrary winds or close quarters, as shown in the Whaling Museum's c 1850 diorama. She was owned by the Peck family, Peck Slip in Manhattan's East River, and also plied Sound waters to New Rochelle and Glen Cove as well as, after 1845, the Hudson River. "A complete list of all the various [ownership] partners would fill an entire page; there were enough of them to just about fill the little sidewheeler to capacity," notes historian Edward Heyl. Today's Eagle Dock Community Beach is the site of her dock.*

An issue of the *Long-Islander* carried her proud advertisement:

"Commencing Friday morning, May 3, 1844, the Splendid Steamboat American Eagle, Capt. Charles Peck, will leave Fulton Market Slip every afternoon at half past 2 O'clock (Sundays excepted) for New Rochelle and Glen Cove. The American Eagle will extend her trips to Oysterbay, Cold Spring and Huntington (Long Island) every Tuesday, Thursday and Saturday. . . . Returning will leave Cold Spring at 6 O'clock, Oysterbay 20 minutes past 6, Glen Cove quarter to 8, New Rochelle quarter past 8, touching at Whitestone and Prime's Dock, each way."

The *American Eagle* earned her keep for the Joneses for many lively years, and was apparently gone from their service by 1849. Among her local legacies is the harbor's Eagle Dock Community Beach, beloved to its swimmers and sailing families today.

The Jones brothers vision that did *not* materialize has to do with railroads. Swift transportation for goods and produce being so vital, the brothers reasoned, why not bring the railroad directly to the harbor's edge? They set about making it happen: a rail link right to the shore. The Articles of Association of the Hicksville & Cold Spring Branch Railroad Company, 1853, confirmed plans for a rail link directly to "the village of Cold Spring at the margin of the water forming the harbor of Cold Spring or as near thereto and as far towards forming a connection with vessels navigating such harbor as shall be practicable." The proposed route? Syosset, then across today's Stillwell Lane and along the western edge of the mill ponds, paralleling Route 108, and on down past St. John's Church to the harbor.

Grading was largely achieved, but the rail link itself did not happen. History records differing reasons why, litigious and otherwise (whaling being on the wane for the Jones brothers . . . squabbles about just where to place the terminal . . . the Jones brothers' deaths in 1855 and 1859). But imagine the potential for industry, population, and substantial physical change if it had. What would have happened to Cold Spring Harbor?

Swift transportation for goods and produce being so vital, the brothers reasoned, why not bring the railroad directly to [Cold Spring] harbor's edge? They set about making it happen. . . .

Whaling ends . . . a new era is in the offing

John H. and Walter R. Jones died within just a few years of each other, not long after mid-century. Their bold little fleet had earlier suffered some devastating losses. The *Richmond* was wrecked ashore in the Bering Strait, 1849, and the *Edgar* grounded and sank in North Pacific waters, 1855. Further, the country's whaling trade was hard hit by the discovery of petroleum in 1859, replacing whale oil virtually overnight, and then by vessels being commandeered or sunk during the Civil War.

Out of the Wilderness records the changing of the times: "And so, with the death of John H. Jones in 1859 and the outbreak of the Civil War immediately afterwards, came the end of the whaling and the cooper's shop, the woolen mills and one of the gristmills and the general store on the western side of the harbor. The only survivors were the general store on the east side of the harbor and the 'new' grist mill." Walter Restored Jones and John Hewlett Jones are buried with other family members near St. John's Church, which the brothers had helped to found two decades before.

End of an era – above left, the CSH Laboratory's Osterhout Cottage, as one of the Bungtown whaling era buildings overlooking a harbor of anchored schooners and old docks and yards, c 1878. . . . John H. and Walter R. Jones had died not long past mid-century after devastating losses – among them the Edgar *(shown at top, with a page from her ship's log), grounded and sunk in icy waters, 1855, and discovery of petroleum in 1859.*

With these two remarkable men ended an era whose legacies we delight to recall. Among them – the thought of whaleships sighting on St. John's white steeple, the Bungtown buildings restored by the Lab, the Whaling Museum's handsomely scripted ships' logs and scrimshaw carvings, and its black-iron signal cannon that once announced the whalers' return from distant seas. Notice the cannon next time you walk along Main Street – a connection in time, just beside the whaleships monument marking the Museum today.

But as the industrious whaling era ended, another, centered around new-found leisure instead, was waiting in the wings. Soon would come elegant excursion steamers out from New York in the 1880s . . . picnics for thousands on our picturesque shores . . . dining and dancing in our elegant shoreside hotels. Soon too would come two institutions whose histories would be remarkable indeed: the Cold Spring Harbor Fish Hatchery in 1883, and the Cold Spring Harbor Laboratory in 1890. One would ultimately offer environmental education and wonder to the families of this town and well beyond; the other would become central to science and to the world.

Water for the Hewlett-Jones gristmill ran from St. John's Pond through a sluiceway straight to the mill, as shown here, then down to power the wheel. The sluiceway can still be seen beneath its dense underbrush, paralleling Harbor Road at the foot of the Route 25A hill.

'OLD MILL' COLD SPRING HARBOR L.I.

Frank Nichols's store, Main Street, c 1900 (today Heritage Candle Shop). Steamfitting and gunsmithing were among the shop's offerings. Whaler Isaac Price is said to have built the original building c 1850.

The old gaff-rigged schooner Margaret Ann *loads wood along Shore Road, c 1915, just as many a vessel did before century's end (sandspit visible in middle distance).*

A favorite end-of-century pastime - fishing at Eagle Dock. A beamy schooner is tied up at left, with Abrams Shipyard dock at right.

Muralist John Banvard's "Glenada Castle," its turreted roofs shown at left within this promotional postcard after enlargement into the grand Glenada Hotel, stood on the hill above today's Cold Spring Harbor Beach Club. The hotel offered many a pleasing pastime during the harbor's steam sidewheeler excursion era. . . . John Totten, Totten livery of Spring Street, is invited to a Glenada Casino "hop," summer 1891.

The Laurelton hotel stood on spacious grounds across the harbor and offered elegant accommodation for summer-excursion visitors. Here, rowing dories, bathing beaches, and a sloop at the dock await the pleasure of guests.

3 Near Century's End: Columbia Grove and Its Elegant Visitors

Today, with the harbor so calm except for tumultuous line squalls and occasional gales . . . a weekend's few unmuffled boat engines . . . and billowy racing sailboats on summer days, it is difficult to picture the water's amazing liveliness just a century ago. These decades – the quarter-century between 1875 and 1900 – were the years of the grand sidewheel steamers. And the 183-foot *General Sedgwick* was among the most beautiful sidewheelers of them all. She made city-to-country excursions here to the spacious Columbia Grove, out near today's Lloyd Neck Causeway. In her work and in her beauty, she exemplified the era.

Whaling was undeniably behind us by now, the Civil War as well, yet memories of both were fresh in many a mind and heart. Coasting sloops and schooners were still plentiful even while being supplanted by steam vessels moving faster, ever faster. Dole's Shipyard, on land that is today part of Eagle Dock Beach, had its hands full in ship salvage work through 1885. Farms and mills still shipped their goods by water. The railroad had reached Syosset well before the Civil War years, just a few miles inland, and Cold Spring Harbor out at Woodbury Road, finally, in 1868. Merchants and bankers could reach their Manhattan offices in record time. And increasingly, Cold Spring Harbor was a destination of the handsomest vessels of the day.

All along the North Shore, starting in these end-of-the-century years, would come a time of elegance that is fascinating to recall. Vast fortunes were being made in New York City; income tax did not yet exist to curb the wealth. Within a generation these fortunes would yield the first of the lavish Gold Coast mansions along the shores of harbor and Sound. Steam-propelled sidewheelers, plentiful and beautiful, offered city dwellers the chance to escape their crowded streets in the heat of summer. Soon, overlooking our picturesque and navigable harbor, three fashionable hotels arose to welcome the excursionists. The hotels were of splendid style: the Glenada, the Forest Lawn, and Laurelton Hall.

Harbor hotels, day trips, and picnics ashore

The Glenada Hotel was built originally in 1853 as artist John Banvard's "Glenada Castle," and later owned and operated as a hotel by the William Gerard family. It stood grandly on the harbor's eastern hillside just above today's Cold Spring Harbor Beach Club; the hotel's 1890 Casino survives handsomely as the Beach Club itself today. The Gerards' Forest Lawn Hotel, completed in 1873, stood just uphill from Shore Road a few hundred feet south – near Eagle Dock.

The third and architecturally most elegant of them all was Laurelton Hall. "Filling its rooms and veranda's rocking chairs to capacity" in 1875, as contemporary accounts noted, the hotel sprawled at water's edge just across the harbor, in today's Laurel Hollow. It had earlier also been managed by the Gerard family, and would later be the site of the Louis Comfort Tiffany estate, completed in 1904 and named Laurelton Hall as well. The estate's smokestack, tall and ornamental, is all that survived when the Tiffany mansion burned in 1957 and is still visible near shore today.

The North Shore's summer hotels met the tastes of a jubilant Victorian-age custom, newly possible as working conditions improved in New York City.

A recollection of the harbor as Laurelton guests must have seen it is captured for us in a familiar work of art: Louis Comfort Tiffany's idealized stained-glass scene of the harbor and Oyster Bay, with lavender-blue wisteria in the foreground. The scene is a favorite in the American Wing of New York's Metropolitan Museum of Art collection.

The North Shore's summer hotels met the tastes of a jubilant Victorian-age custom, newly possible as working conditions improved in New York City. It was a custom enjoyed by the richest and the far less rich as well: leisurely steam sidewheeler outings along the Sound . . . Sunday picnics in Columbia Grove and elsewhere around the harbor . . . dressy dining and dancing ashore in the harbor hotels. Church groups, singing groups, families escaping the city's ways – they all came to Cold Spring Harbor and neighboring North Shore towns. Hempstead Harbor, Sea Cliff, Oyster Bay, Northport, and Port Jefferson were among their many destinations. In local response, our three hotels arose, boarding houses opened, and farming families took in summer boarders. Main Street's Burr-Gardiner Hotel, near Shore Road, was among the establishments welcoming summer excursion visitors in these years.

Grand and graceful steam sidewheelers attracted thousands of passengers at a time for their excursions. Families came ashore for festive picnics at Laurelton Grove. On the eastern shore

Forest Lawn Hotel, completed in 1873, ranged up the Shore Road hillside across the street from Eagle Dock. . . . This Gerard family gathering – out on the lawn overlooking the harbor, with tea – was clearly a pleasing occasion.

they debarked at Eagle Dock, at Wawepek Grove near today's Lloyd Harbor Beach, and at Columbia Grove near the Lloyd Neck Causeway. On return trips, to the mutual satisfaction of ship owner and local farmer, the steamers carried freight and fresh farm produce down the East River to Manhattan markets.

Soundings gives us colorful accounts of the era: "Day after summer day, large two- or three-decked side wheeler steamboats, towing usually one or more awninged barges, plied up and down the Sound bringing crowds of excursionists to enjoy the delights of seashore and countryside at small cost." A summer 1879 advertisement for a day-long outing in the "splendid steamer *General Sedgwick* . . . music by Wagner" offers the Cunard / East River-Columbia Grove trip for $2, "admitting gentleman and two ladies . . . extra ladies tickets 50 cents." City advertisements for the Peck Slip-Cold Spring Harbor run, offered by the Oyster Bay and Cold Spring Harbor Transportation Company, state the fare as 40 cents.

Cold Spring Harbor was particularly popular among the string of North Shore day-trip destinations in these years. Festive passengers poured ashore to relax in picnic groves and hotels . . .

Cold Spring Harbor was particularly popular among the string of North Shore day-trip destinations in these years. Festive passengers poured ashore to relax in picnic groves and hotels, responding to effusive promotional literature:

" 'The [Glenada] hotel is on high ground surrounded by acres of woodland and well-kept shady lawns sloping gradually to the waters edge.' Here a salt marsh had been changed into a park-like area, a two-story casino built . . . lawns and tennis courts laid out. . . . 'The Casino was fitted with a ballroom, cafe, ladies billiard room, local and long distance telephone and Western Union Telegraph office for the exclusive use of Glenada patrons.' The patrons 'are assured of amusements of great variety, boating, bathing, sailing, fishing, tennis courts and croquet grounds and excellent roads for riding, driving or wheeling.' "

Across the harbor, "The Iron Pier Steamboat Company served all Laurelton excursions, managing the grove and providing the recreational facilities of bath houses, benches, and tables, swings, carousel and a dance pavilion where the cotillion bands and stringed quartettes that played for the shipboard dancing, continued their programs of polkas, gavottes, mazurkas, gallops, lanciers, quadrilles and waltzes during the hours at the grove. Food and soft drink concessions were let to local bidders."

Excursion vessels of the day included the 1840 steam sidewheeler *Croton* as well as the newer *Idlewild* and *Shadyside*. And, her graceful lines captured by Edward Lange in an 1881

The 179-foot steam sidewheeler Idlewild, *built in Brooklyn in 1876, joins canal boats, sloops, schooners, steam tugs – and at least one square-rigger – just below the newly opened Brooklyn Bridge.* Idlewild *was familiar on the Peck Slip-Cold Spring run during the excursion era.*

11th ANNUAL

EXCURSION

OF

Hiram Lodge

No. 17, F. & A. M., to Columbia Grove, Long Island, on

THURSDAY, AUGUST 7th, 1879.

Leave Cunard Dock at 8:30 A. M.

The splendid steamer General Sedgwick, and commodious barges have been secured. Music by Wagner, of this city. Every effort will be made to ensure a pleasant and enjoyable time.

TICKETS, $2.

Admitting Gentleman and two Ladies. Ex'ra ladies' tickets, 50 cents.   jy29 w.s.m,tu.w 1p

A common harbor sight from the mid-1860s to 1875, the graceful 207-foot sidewheeler D. R. Martin *(shown here with her later name* Howard Carroll*) carried passengers from city to Sound - including Oyster Bay and Cold Spring Harbor - on regular runs for the Northport, Huntington, & Oyster Bay Steamboat Company. An advertisement offering her for sale described her with "ample room for one thousand people . . . drawing not over five feet with fuel on board." As "D. R." she was affectionately known as "Dr. Martin" by her passengers. . . . Members of Manhattan's Hiram Lodge are invited to take a day-trip to Columbia Grove, Long Island, on the "splendid steamer General Sedgwick" in August, 1879. Ticket price, $2 for a Gentleman and two Ladies.*

COLUMBIA GROVE.

LLOYD'S NECK. L.I.   GEO. VAN AUSDALL, PROPRIETOR.

The General Sedgwick *at Cold Spring Harbor's Columbia Grove, her graceful lines captured by artist Edward Lange in 1881. Picnics here, and at other groves around the harbor, were powerful attractions for church groups and other city dwellers in these end-of-century years.*

Cold Spring Harbor scene, the elegant *General Sedgwick* at Columbia Grove. Alongshore, Lange's families swim at harbor's edge and go rowing in the shallows. Parasoled ladies stroll along in ruffled dresses, with men in dark summer suits and hats. Mothers and children stop to talk. Open carriages with fine matched teams of horses head swiftly north and south along the road to Lloyd Neck. It's a grand scene of bygone days, and today many of us are surprised to know it ever happened at all.

Smaller vessels moored in the harbor as well, in those years, often with style that rivaled even the finest sidewheelers. On summer weekends, season after season, luxurious yachts sailed in to anchor in beautiful surroundings, their passengers coming ashore for festivities at Laurelton or Glenada. Annual yacht club cruises gathered in the harbor's sheltered anchorages – graceful sloops and yawls and ketches of the finest design. Oyster Bay's Seawanhaka Yacht Club was founded in 1871, conveniently east of Hell Gate for the wealthy Manhattan residents summering here. The Cold Spring Harbor lighthouse was built in 1890, undoubtedly in response to the comings and goings of the working sloops and schooners, the roomy steam

Cold Spring Light House, Cold Spring, L I,

My Dear Mattie how are you this morning be careful how you go on the ice & tell mama to get her dress braid when she goes to Burlington Ga

Cold Spring Harbor's 1890 lighthouse, marking the sandbar that deep-draft vessels must skirt. Sailors still tack around its gargantuan black base today; the lighthouse structure stands quietly ashore on private property. . . . Excursion-era elegance: watching the Seawanhaka Cup just off Lloyd Neck, c 1880.

The 129-ton schooner Albert Pharo *is licensed by Act of Congress to serve in the coasting trade, 1873, by Jacob C. Hewlett, Surveyor, in continuance of Cold Spring Harbor's Port of Delivery designation of 1799. . . . Steamer* Port Chester *bill, April 1894, for "use of pony, 35 days." The vessel was operated by the Oyster Bay and Cold Spring Transportation Company on New York–Centre Island–Lloyd's Dock runs: "Freight horses and carriages handled at reasonable rates . . . passenger fare 40 cents."*

At Cold Spring Harbor station, 1881, families enjoy the swings, benches, refreshments, and cooling shade of Woodbury Park, reflecting the station's early designation as Woodbury Station. Friends wave across the tracks as the train for New York chugs in.

sidewheelers, and the teak-trimmed yachts of these grand and prosperous years.

By no means was the elegance of the era limited to ships and yachts, however; the railroad had at last come into its own. Once the all-rail link to Boston had been completed by the competition along the Connecticut shore, in 1848, the Long Island Rail Road shifted its attentions to the Long Island towns past which it ran. Service had reached Cold Spring Harbor at last by 1868, after complex and feisty negotiations among several vying lines. But service in and out of New York was eventually most accommodating and comfortable, and in 1884 local promotional literature extolled the comforts of travel by rail:

"The arrangements for the transportation of passengers with promptness, speed and comfort are equal to those of any railroad in the country . . . making it feasible and agreeable for a

CSH firemen with horse-drawn firetruck, c 1890; the red-shingled first firehouse, shown here, is tucked behind the modern facility today.

West Side School class of 1889 poses for a picture; mufflers, full skirts, knickers, outgrown coats, and ankle-high boots reveal style and season. Among meticulous records kept by West Side are Minutes books from the 1790s, and a commemorative publication noting that Miss Georgia D. Titus, teacher and later principal, received $600 in salary in 1876. . . . Miss Titus relaxes on her porch overlooking the harbor, August 1892.

business man to be at his countinghouse in town during the day, and to reach his summer home before nightfall. . . . The celebrated Woodruff parlor . . . provides the very acme of comfort . . . insuring noiseless and gentle motion . . . perfect ventilation is secured without drafts, and even in the warmest weather the cars are delightfully cool."

Grand harbor-side hotels, family excursions and weekend picnics in popular groves ashore, rail trips in handsomely appointed cars, women with parasols and swingy long skirts and men in striped jackets and bowlers – these were grand and prosperous years that are all but unknown to us today. Fortunately, however, a full and fascinating account of the harbor's resort years is available in the archives of the Village of Lloyd Harbor, awaiting publication in the near future.

"There was the fun of Spring freshets when Estelle and I 'dammed up' the brook. Pieces of bark served as our ships. Mine would be named the William L. Peck *after my father's clipper schooner."*

Highlights of an era

While the grand steam sidewheelers were familiar Cold Spring Harbor sights in the 1880s and 1890s, so also were the harbor's working sloops and schooners, and the shipyards and farmhouses ashore. Mildred E. Bunce, the granddaughter of local ship captain Joseph Titus Bunce, remembers the late 1880s winters in Cold Spring Harbor in a November 1966 issue of *Long Island Forum*. Her grandfather's farm was on Goose Hill Road just east of the village; the handsome house still stands today:

"Clothes were brought in from the line . . . frozen into effigies. . . . They smelled of absolute cleanliness." Girls warmed their nightgowns at the downstairs fireplace before going up to bed. Of the famous Blizzard of 1888 Mildred writes: "The wind suddenly rose and whipped the sturdy maples in the yard in to topsy turvey frenzy. The live stock fidgeted. . . . The wind wailed through the halls, up the chimneys with the eerie sound of a banshee. Windows tugged at their hasps, doors chittered. Terrorized birds wheeled and flung themselves against the wind. . . . This was such a terrific snow storm that the children thought it might be the end of the world."

And of one Christmas time she recalls: "I found a Red Riding Hood doll and a toy xylophone under the tree for me. They were so precious to me. The next morning . . . I walked the length of Goose Hill Road looking for reindeer tracks." And "Who can forget the zing of skates on ice? We skated 'round and 'round the duck ponds."

Of springtime with a young friend, Mildred Bunce remembers: "There was the fun of

Catboats, dories, and a big schooner out at the sandspit keep company with clam diggers at low tide, c 1900.

Commodore William Vanderbilt's North Star, *in her heyday. Years later the vessel, her illustrious career complete, lit up the skies as she was burned for scrap at Dole's Shipyard, Cold Spring Harbor.*

THE NORTH STAR.

Spring freshets when Estelle and I 'dammed up' the brook. Pieces of bark served as our ships. Mine would be named the *William L. Peck* after my father's clipper schooner. Estelle would name her ship *Export* for my grandfather Bunce's ship. He was Capt. Joseph Titus Bunce Sr., my father being Jr."

Within the fuller context of this summer resort era of stylishness and change – the era in which young Mildred Bunce lived – three other Cold Spring Harbor events shed particular light. The first event, the burning of Commodore William Vanderbilt's steam yacht *North Star* for scrap in the mid-1860s, predated the excursion steamer era but shows an aspect of the work of Dole's Shipyard through 1885.

Commodore Vanderbilt's elegant 270-foot ocean steamer *North Star* had been well known in her day – cruising to Europe in 1853, then carrying passengers back and forth to Central

America's cross-Panama train for the trip up along the California coast. She was an absolutely beautiful ship. He was among the richest men of his day. Among his financial triumphs – the Staten Island ferry and the exceptionally successful New York Central Railroad. The *North Star* had completed her illustrious career by serving as a troopship in the Civil War, then suffering a string of mishaps including catastrophic hurricane damage off Cape Hatteras. She was now up on the sandbar south of Eagle Dock, ready to be burned for scrap.

Helen Rogers, our town's mid-century diarist, describes the night in a later written account. Her words seem like a Fourth of July flare, briefly revealing the surrounding harbor scene:

"Those who were in the vicinity of Cold Spring Harbor on the night of the 10th of October [1865] had the rare opportunity of witnessing the splendid spectacle of a ship on fire without the sad associations usually connected with such a sight. The great steamship which had been beached for sometime, being dismantled and stript of her sheathing, was on Tuesday consigned to the flames for the sake of her metals and altho' without mast, pipes or wheelhouse, yet lying on a beach far down the harbor, surrounded by dark waters. The appearance at night was magnificent. The beautiful bay with its wooded heights, winding shore studded with pleasant cottages and country seats and . . . the beach itself, strewn for acres with piles of copper, iron doors, and great water tanks, boilers and pumps and trans-Atlantic life boats. The other vessels were hulks looming up in the distance – desolate, waiting for the torch . . . all lit-up by the great sheets of flames which blew out phosporous lights that played through the yawning crevices, shown like rockets through columns of rolling dark smoke as glowing timbers fell in the water. They gave forth hissing sounds which sounded fearfully in the stillness of the night. So perished the *North Star*, the most celebrated steamship of her time."

The second excursion-era vignette is just an ordinary outing by inquisitive schoolboys. It is typical of its time and, like the *North Star's* burning, concerns salvage activities at Dole's Shipyard. Such activities took place throughout the 1860s-1880s; the busy harbor that they served must surely have shown many a wonderful afternoon to schoolmates out on a lark. Harriet Valentine, in *The Window to the Street*, quotes an April 1877 entry in the diary of Harry H. Funnell, one of the adventurous boys:

"We waited a good while to see an old frigate towed into the harbor to be torn up and burned. . . . The [boys] got up a party at school to go over and visit the frigate. . . . There was only a ladder and some ropes and cleats to climb by, but we all got on and had a fine time hunt-

Stoots Dairy, uphill from Main Street, c 1900. Diarist Helen Rogers's house, "the Vineyard," is visible in the distance. . . . Totten livery in an 1881 watercolor gouache by artist Edward Lange. The stables stood on Spring Street near Shore Road, just behind Main Street and a block from the Rogers house. Remnants of a split rail fence mark the site today.

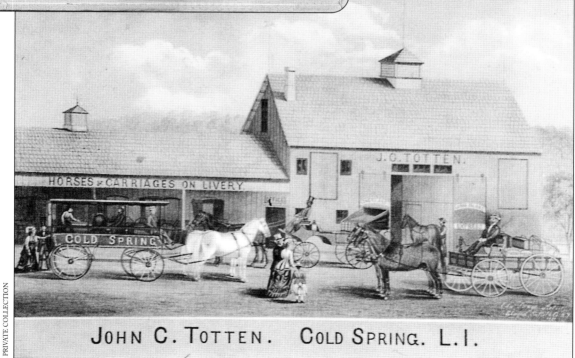

43

Jones Hill (Route 25A/Fish Hatchery Hill today), shown here sometime after 1907, was a challenge in winter. . . . In warmer weather, carriage rides along Shore Road (heading north, Eagle Dock in background) were pleasant outings.

ing around above and below deck." Picture the frigate in a harbor lined with old brickyards, docks, and mills, with sloops, schooners, and steamships all carrying their cargoes back and forth between Sound and harbor dock. What a great afternoon's outing *that* must have been!

It was in this busy Cold Spring Harbor, its business section full of trade and bustle and its

The excursion barge Republic, *built in Poughkeepsie in 1865, was rebuilt and renamed* Columbia *after the August 1891 disaster in Cold Spring Harbor.*

waters and shores attracting droves of summer Sunday picnickers, that the third event took place in mid-August 1891. It was a tragedy that would make headlines around the world today.

The *Long-Islander* records that the 167-foot excursion steamer *Crystal Stream*, towing the beamy double-decked barge *Republic* alongside, was routinely docked out on the eastern shore at Wawepek Grove. Together the vessels carried hundreds and hundreds of passengers, many of whom had streamed ashore for swimming and picnicking as was the custom of the day. Others relaxed aboard in the cooling shade of the barge's lower deck.

A sudden squall came up – bringing gale force winds and, undoubtedly, the familiar black skies and hissing-white torrents of rain sweeping sideways across the Sound. Families ran for the shelter of ship and barge, crowding on board. In the violence of the winds, news accounts note, the upper deck of the barge lifted from its stanchions and collapsed down onto the deck below: "The terrifying cracking, splintering, crashing sound like the noise of the fall of a huge forest tree, was heard by those on shore above the tumult of the storm." Hundreds were injured, and twelve passengers died. Cots set up in the Glenada Casino's bowling lanes served as makeshift hospital space. According to shipping news accounts in Manhattan, notes maritime historian Norman Brouwer, the incident occasioned long-needed new standards and inspections for the excursion steamers and barges of the day.

The tragic squall and the *Crystal Stream* . . . Mildred Bunce sailing tiny ships in the spring freshets . . . schoolboys exploring a derelict frigate towed in for scrap . . . the *North Star* lighting up the harbor as she burns . . . picnics in Columbia Grove and elegant dining ashore in the Glenada Hotel: each small story illuminates an era long past. Each centers on this particular harbor, connecting us to earlier generations, strengthening our sense of this place. Likewise the harbor's seasonal squalls, nor'easters, and choppy grey days that came then just as they come now. And likewise, the sweet afternoon southerlies and exquisitely clear shimmery days in Cold Spring that are so like those of other harbors, in other times.

Each small story . . . centers on this particular harbor, connecting us to earlier generations, strengthening our sense of this place.

Two institutions begin their histories: the Hatchery and the Lab

It was in this same quarter century of the town's history – the years of schooners and yachts, sidewheel steamers, and grand hotels – that two remarkable institutions came into being: the Cold Spring Harbor Fish Hatchery & Aquarium in 1883, and the Cold Spring Harbor Laboratory in 1890. Their histories link us clearly to our past.

The new New York State Fish Hatchery (which became today's privately run Fish Hatchery & Aquarium in 1982) began its work in a former Jones family woolen mill, a small brown-shingled structure just north of Jones Corner and the foot of the hill. The Hatchery was the state's very successful effort to supplement the natural spawning of trout in its southeastern lakes and streams. New York Fisheries Commissioner Eugene Blackford, long influential and active in the Fulton Fish Market in New York, was its founder. Its first manager was noted fish culturist and author Frederic Mather.

Hatchery documents confirm that the site was chosen advisedly. Notes an 1880 Fisheries Commission report: "Long Island is an island of springs and spring brooks . . . there is something, too, in the temperature character of the water that is favorable to trout, for the young even when kept in confinement and fed on natural food, grow nearly twice as fast as they do in many of the inland streams and ponds of the State."

Jones family members, Whaling Company descendants, helped immeasurably in the founding. The help went well beyond making the old mill property available to the state. When stone

class of the Biological Laboratory. 1890
Returning from a field trip.

The harbor and village in which the Fish Hatchery and the Cold Spring Harbor Laboratory began: salt meadows at the head of the harbor . . . the former Jones woolen mill in which the Hatchery started in 1883. . . Summer's Biological Laboratory class members exploring the intertidal shallows in the naphtha launch Rotifer.

was needed for the salt water pond's big retaining wall, for instance, Townsend Jones sent three schooners across the Sound to haul it from Connecticut's copious brownstone quarries. Further, when the time came for a superintendent's house, Townsend and his brothers John and William got together and arranged for "a large and elegant dwelling" of nineteen rooms to be built just above the harbor. The residence was notably handsome in its time. It was beautifully restored by the Cold Spring Harbor Laboratory in 1980 as Davenport House, rich in its original dark green and pumpkin hues, and stands just uphill across Route 25A from the Hatchery today.

Attention from the Jones family would remain unfailing. Twenty-five years later in 1904, as reported in the *East Norwich Enterprise* of March 5, there came a catastrophe: "The heavy rain of Sunday caused the dam at St. John's Lake, Cold Spring, to burst, which carried away nearly 100 feet of the roadbed, flooding the grounds of the State Fish Hatchery and doing great damage. . . . The main highway from Cold Spring Harbor and Huntington to Oyster Bay is now impassable. . . . The lake . . . is entirely dry, the fish having been washed into the harbor."

The Hatchery and its near neighbor the Cold Spring Harbor Laboratory still enjoy the interconnection, century-long, of their histories.

But by a year later, the paper reports, members of the Jones family had restored "a tumbling dam over which the water continually pours in a succession of cascades. It makes a pleasing picture and attracts the admiration of all who pass that way. . . . St. John's Lake is to-day as beautiful as ever at this time of year." And as visitors observe today in the 1990s, some things never change. The Hatchery's 1889 annual report recorded "young visitors catching crickets, and throwing them in the ponds to watch the flashing splash of the trout."

The Hatchery and its near neighbor the Cold Spring Harbor Laboratory still enjoy the interconnection, century-long, of their histories. In her book *Houses for Science* author Elizabeth L. Watson notes an early instance: "The establishment of the Fish Hatchery at Cold Spring Harbor was to have unexpected consequences for this quiet village. Fish Commissioner Eugene Blackford also sat on the board of the Brooklyn Institute of Arts and Sciences and was well aware that fellow board member Adelphi University zoology professor Franklin Hooper (1851-1914) was eager to have the Brooklyn Institute set up a seaside zoological station. Talk of Darwin's theory of evolution was in the air, and it was the latest vogue in zoological circles to study nature at its source – the sea."

The Lab's first biology classes met at the Hatchery in 1890. As the fledgling institution

Route 25A causeway under construction, c 1907, with the old milldam path in foreground, Hewlett-Jones gristmill at upper right. . . . The marshlands and the old Hewlett-Jones gristmill at low tide.

moved into quarters across the road on the harbor's southwestern side, new life came to the buildings once bustling in whaling days – the days of the Bungtown coopers, shipworks, mills, and sail lofts. Its first schoolhouse and lab were completed in 1893 and could accommodate sixty students, salt and freshwater aquariums for the study of sea and intertidal creatures, and private labs for six visiting scientists. Today the Lab is home to more than two hundred scientists and is respected worldwide for its cancer and genetics research, its symposiums attracting and interconnecting brilliant researchers from all over the world, and its association with several Nobel laureates, including Lab president and former director James D. Watson (who shared the 1962 Nobel Prize for discovering the structure of DNA), as well as Alfred Hershey, Barbara McClintock, and Richard Roberts.

In the century and some since its founding, the Lab has taken its Bungtown and other histories seriously indeed. Archaeological site reports document the Matinecock, colonial, and whaling eras here. Arrowheads, kaolin pipes, pottery, shoe lasts, and harpoon fittings are among the artifacts that tell the story, and, now as part of the Whaling Museum collection, enhance the Museum's ongoing local history interpretation.

The Lab's seven pre-whaling and later nineteenth-century buildings (among them the 1805 Airslie House built by Major William Jones, the whaling era residence Hooper House and warehouse Wawepex Building, both c 1850, and Davenport House built in 1884 for the Fish Hatchery) are ranged along the historic Bungtown Road. Each of the seven has been meticulously restored for its new career in science.

Cold Spring Harbor at century's end

The nineteenth century was ending with a flourish. The Library, Post Office, Fire Department, and other essential local institutions were growing to meet new needs. Each week, then as now, the *Long-Islander* continued to document the details of these years. Cold Spring Harbor kept proud pace with other Long Island communities, and the times were seen to be good.

Among the most apt and insightful comments in Newsday's new book *Long Island: Our Story* is a scene-setting one by staff writer Steve Wick (reprinted with permission © Newsday, Inc., 1998). The article quotes SUNY Stony Brook historian Roger Wunderlich as saying that the past was still very much with us as the nineteenth century was ending, and Civil War veterans still marched proudly in our Fourth of July parades. But whaling and its international connections were gone, and as Wick observes, "New York's exploding population was pushing east like a modern-day glacier, reshaping the landscape all over again." By century's end, the article notes, good railroad connections enticed residents out from the city all summer long, and the Gold Coast era of mansions and elegant lifestyle was metamorphosing the North Shore.

· · ·

This harbor of Cold Spring, with the grand shoreside mansions such as Walter Jennings' *Burrwood*, Louis Comfort Tiffany's *Laurelton Hall*, and Robert de Forest's *Wawapek*, was part of it all. Here as in the world, the early twentieth century did indeed bring undreamt-of changes – as has been so beautifully documented in the books and exhibits of many a Long Island historical society and museum. Today at our own turn-of-the-century time – only a few generations

From old century to new, c 1901, with a gaff-rigged schooner ghosting in to Abrams Shipyard, and a big schooner up on the ways ashore. The Glenada Casino is at center, with cupola-topped Eagle Dock just south.

beyond the summer resort years here – the place thrives in its schools and churches, its post office and Main Street shops, all metamorphosed to modern ways but all meeting the community's changing needs just as before.

Cold Spring Harbor's local institutions reflect our history and lives today, as do our generation's own wonders (among them television and space travel, computers and cellphones, and our global and digital interconnections regarding so many matters large and small), which would have so astounded our predecessors here. But in these new dimensions lies the invitation always issued by time, and responded to so revealingly in the written and oral histories already so treasured by this town. The early twentieth century, our own turn-of-the-century opportunity, is still very much in the memories of our elders. Perhaps, as we walk along Cold Spring Harbor streets and shores today, and delight in the details of generations and centuries ago, we would be wise to respond to the invitation. To ask questions, to listen, to note the evidences of what has gone before. To keep journals and otherwise document our own times as well.

For tomorrow's history truly lies in the ordinary details of our days.

Cold Spring Harbor
Timeline Highlights ❷

Chapter 1 –

18,000 years ago – glaciers
form North Shore hills
2500 BC Early Matinecock
presence on Long Island
1653 First Purchase

Chapter 2 –

1681 First mill, Upper Pond
1782 Conklin paper mill
1790 West Side School
1836 St John's Church
1836-62 Whaling years
1868 Railroad to CSH

Chapter 3 –

1875-1900 Summer resort/
steam-sidewheeler years
1883 CSH Fish Hatchery
1890 CSH Laboratory
1890 CSH Lighthouse

. . .

Chapter 4 –
After 1850

1. Glenada,
Forest Lawn hotels
2. Laurelton Hall hotel
3. Whaling Museum,
DNA Center, SPLIA
4. Fish Hatchery
5. Davenport House/
CSH Laboratory
---- Route 25A, 1907

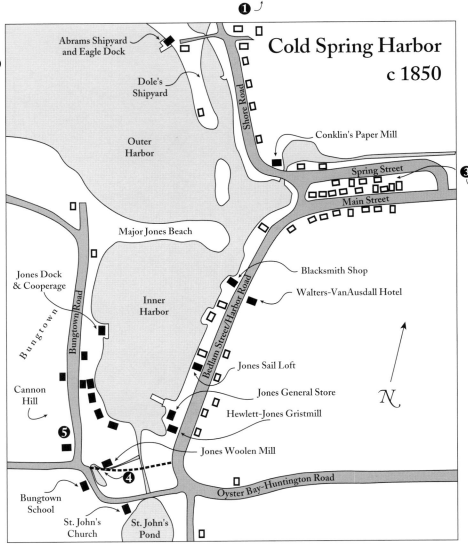

Cold Spring Harbor
c 1850

Abrams Shipyard
and Eagle Dock

Dole's
Shipyard

Conklin's Paper Mill

Outer
Harbor

Spring Street

Main Street

Major Jones Beach

Blacksmith Shop

Walters-VanAusdall Hotel

Jones Dock
& Cooperage

Inner
Harbor

Bedlam Street/Harbor Road

Jones Sail Loft

Bungtown Road

Jones General Store

Cannon
Hill

Hewlett-Jones Gristmill

❺

Jones Woolen Mill

N

Bungtown
School

❹

Oyster Bay-Huntington Road

St. John's
Church

St. John's
Pond

4 The Harbor's Streets and Shores Today

Cold Spring Harbor is one of those rare communities where history is apparent from the moment a visitor makes the sweeping turn to head down Fish Hatchery Hill. Rich-hued Davenport House is on the left, St John's Church and the Fish Hatchery on the right, and, just as the phragmites marshlands begin, the road bends left to hug the eastern shore of the inner harbor. Everything changes here – glancing light bounces off ruffled water, old shingled houses with windows of shimmering glass gleam out of the shadows on either side of the road, a gull cries, sea-salt sharpens the air. No question about it: this is a maritime place, one that has been tended and treasured by its people for a long stretch of time.

There is an art to understanding the imprint the past has left on the present. To do it well takes an attentive eye and a willingness to look both at the surfaces, and into the shadowy spots that are sheltered by them. Historic places reflect all the times and the events that formed them, but some of these have left stronger prints on the land, and in people's minds, than others. Sometimes, whole eras virtually disappear in the stronger traces of more prosperous – or seemingly more glamorous – ones. This is true of Cold Spring Harbor, where the proud focus on whaling days overshadows subtler evidence of the subsequent era when resort hotels and steamboat tourism flourished. It was during this later time that much of the town as we see it today was retrimmed, or more

MAIN STREET SKETCHES BY ANNA DAM-VOLKLE

Today's Main Street, Cold Spring Harbor.

substantially changed. It is possible, though, to read notations from the entire record, if one's eye is properly sharpened and informed.

To trace Cold Spring Harbor's history through the streets and shores and marshes of the present, the *Whaling Museum*, built at the eastern end of the street in 1942, is a good place to start. Just in front stands the Whalers' Monument boulder, a relic of prehistory given added meaning by its slate plaque commemorating the 1836-1862 heyday of the local whale fisheries. The anchor and capstan nearby are symbols of the maritime nature of much of the story that is about to unfold. But what's a cannon doing in front of the Whaling Museum? A look at the Museum's diorama inside, showing the harbor in about 1850, reminds us that in the whaling years the cannon stood atop a hill on the west side of the harbor. It was fired upon the first sighting of a homebound vessel.

The diorama and its map suggest a good route – first Main Street's Captain's Row, then west from Turkey Lane and west again from Spring Street nearing the harbor, and finally the Spring Street-Shore Road houses and then Harbor and Bungtown roads along the harbor's eastern and western shores.

Just as the whaling years of 1836-1862 blended into the years of the Civil War, this eastern end of Cold Spring Harbor's Main Street was transformed

Harbor diorama, pp iv, 22

Harbor map, p 52

Whaling era, p 17

54

CONTEMPORARY PHOTOGRAPHS BY
JOHN C. STEVENSON

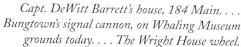

*Capt. DeWitt Barrett's house, 184 Main. . . .
Bungtown's signal cannon, on Whaling Museum
grounds today. . . . The Wright House wheel.*

by maritime fortunes and skilled carpenter-builders into an elegant *Captains' Row*. The Row was buffered from the bedlam of "downtown" to the west by a fringe of the smaller, still elegant, residences built for seamen of lesser authority and practitioners of other, upland trades. Just west of the Museum, and housing its administrative offices, stands the late-nineteenth-century *Captain James Wright house*. The ship's wheel centering the trim at the porch gable field replaces an original said to have come from a coastwise trading vessel Wright commanded. The Museum itself stands within the site once known as *Titus Grove*, a favorite nineteenth-century spot for picnics and strawberry socials.

Three houses, on both sides side of the street near the Museum, introduce a style and a catalogue of architectural features that appear again and again on this street and elsewhere in town. *Captain DeWitt Barrett's house* (184 Main Street), his sister's house (215 Main), and a third just to the east (today a nursery) bear a strong family resemblance. With their three-bay-wide, center-entry plans, and their steeply pitched front gables breaking out over the cornices to enliven the roofs, these houses are classic Cold Spring Harbor architecture of the mid-1860s. They are an exuberant mixture of the Victorian Italianate and Gothic Revival styles whose design elements contribute greatly to this town's distinct sense of place. Whaling captain DeWitt Barrett's story illuminates a peculiar aspect of local history: no native residents commanded the vessels of the local whaling fleet. All were too inexperienced to do so when whaling started here, although several, including Barrett, later rose to command responsibilities on vessels sailing from other ports.

Heading west from Turkey Lane

Just past Turkey Lane, a group of south-side houses draws the eye to bold, Italianate-style details. Some of them have stories as evocative of the past as their architecture. The *Captain Manuel Enos house* (208 Main), with its pink, louvered shutters, was built around 1869 for a native of the Portuguese Azores who, decades earlier, had served in the Cold Spring Harbor whaling fleet. The shutters evoke the pink-trimmed white of the Azorean whaleboats. Enos came to Cold Spring from Sag Harbor, serving aboard the Cold Spring whalers *Huntsville* and *Sheffield*. Later, married to a local woman named Susan Brush, he twice retired from the sea and tried valiantly to build a new life on shore. Nothing he attempted succeeded. So the old seaman returned to whaling in its waning years only to vanish in the bark *Matilda Sears,* presumably lost at sea, shortly after this elegant house was finished.

The *Moses Rogers house* (198 Main), said to have been built by 1840, bears evidence of a later nineteenth-century remodeling in the strong upward thrust of the wood finials crowning the apex of its roof gables, and two-over-two light window sash. Moses Rogers is the man who, with Richard Conklin in the 1820s, divided Main Street building lots from their own lands, setting the stage for residential development in the years to come.

The *Union Baptist Meeting House* (170 Main Street) stands at the corner of Poplar Place. This clapboard chapel, completed in 1847 in a distinctive local interpretation of the Greek Revival style, cost $850 new. It was one of three local churches in the whaling era, and it featured special sermons for embarking sailors, as well as the harbor baptisms at which mischievous local boys urged extra dunkings.

West of Poplar Place, a wooded hill rises steeply over the south side of Main Street. Three small houses (148-156 Main), built into the side of the hill beyond the corner, are among those said to have been framed with timbers salvaged at local shipyards. While this remains a matter for conjecture,

Capt. DeWitt Barrett, p 18
Capt. Manuel Enos, p 23
Building Main Street, p 11

Union Baptist Meeting House, 170 Main.

Heading west from Turkey Lane

At left, decorative finial atop Moses Rogers house, 206 Main.

Capt. Manuel Enos house, 208 Main.

the age of the houses can be read in the dimpled shimmer of the window glass in their low, three-light attic windows.

Just beyond Poplar on the other side of Main, the dogleg extension of Spring Street runs off to the north. Deep within a yard that at various times past was shaded by grape arbors or crackly with silkworms chewing mulberry leaves, the *Daniel Rogers house* (6 Spring) faces Spring Street. It was named *the Vineyard* for the end product of the grapes, sometimes shipped in to New York City markets. This is the house of young diarist Helen Rogers, and was built for her father in 1826. It was from this house that Helen watched ships be built, and, at rainy-day high tides, the flooding of Spring Street end to end.

Looking west from Spring Street

West of Spring Street commerce begins. The shops and stores and taverns and houses on the west end of the street have coexisted in mixed use since the beginnings of the village. This is the *Bedlam Street* of whaling days. . . .

The *Velsor* (181 Main) and *Scofield* (169 Main) houses are virtually identical, both having combined residential and commercial uses for most of their histories. They were built in stages between 1820 and 1850 in the Greek Revival style, and remodeled in the 1860s. Jacob van Velsor, a butcher, lived above his ground-floor business; 181 Main was owned by Fish Hatchery superintendent Charles Walter in later years. The Scofield house was the Methodist parsonage in the 1870s.

The *Methodist Episcopal Church* (161 Main) was built in a Gothicized version of the Greek Revival style in 1842. Although the original double and triple pointed-arch windows on the bell tower's front can no longer be seen, the similarly pointed arches marching along the eastern side wall give away the structure's stylistic secret. Shipbuilders Moses Rogers and Israel Valentine, and paper mill owner Richard Conklin, were among its prominent members. This handsome clapboard church is today the headquarters of the Society for the Preservation of Long Island Antiquities (SPLIA).

The Federal-style *commercial building* (117-123 Main), displaying the nameboard of the ship *James Cook* high up on its facade, bears visible witness to a practice that is usually dismissed as architectural legend. It is at least sensible rumor that many Cold Spring Harbor buildings utilized ships' timbers, salvaged

Helen Rogers diary, p 21

Bedlam Street, p 9

Methodist Church, p 14

Looking west
from Spring Street

Scofield House and detail of Velsor House, 169 and 181 Main. . . .
SPLIA headquarters at Methodist Episcopal Church, 161 Main. . . .
James Cook *nameboard at 117–123 Main.*

at John Dole's Shipyard and elsewhere, for framing. This one apparently yielded proof: the *James Cook* nameboard found, carpenters say, in an integral position among timbers being examined during a restoration.

Nearing the harbor

On the south side of Main, at the sign of *Bedlam Street* (90-94 Main), the Holmes bakery once made pies and cakes for the townspeople, and hardtack for ships' stores. These heavy crackers were packed in barrels made across the harbor in Bungtown. Due to entomological impurities in mid-nineteenth-century flour, the sailors always rapped their sea biscuits on a hard surface before taking the first bite.

A few doors down, the c 1830 *Valentine-Denton house* (60 Main) with its unusual two-story porch tells of early days and town-building in the arrangement of its floor levels. This house is older than any of its nearby, south-side neighbors. Built into the side of the hill, the walls are much higher above grade on the street side than in the back. The ground may well have been cut down in front of this house, for street-grading purposes, between the time it was built and the late nineteenth century. Like most of its neighbors, this building has simultaneously been both house and store. For a time late in the century, Maggie Brown ran a bakery "at the sign of the bread loaf" on the bottom floor. Her husband Frederick, a ship's carpenter, commanded the coastal schooner *Tunis Bordine* in the 1870s, and supplied departing ships with Cold Spring Harbor's legendary "sweet, potable" spring water.

Across the street on the north side, the handsome clapboard *Conklin House,* or *Seaman's Railroad House* (75 Main), is one of the oldest buildings surviving in town. Said to have been built c 1720, this was by the late eighteenth century the home of Richard Conklin, whose paper mill stood nearby. During the whaling years, in the 1850s, Ezra Seaman operated a hotel here for sailors and travelers arriving by stage from the railroad station in Syosset. The low "frieze windows" on the top floor lighted the tiny, cheapest rooms where ordinary seamen lodged. There is a persistent story that a brothel flourished on the upper floors of this house during the hotel years. On the second floor, up a Federal-style staircase of surpassing delicacy and grace, the madam ran her operation from a large corner room with a fireplace.

A few more Main Street structures reward a glance. A *double store*

John Dole's Shipyard, p 41

Bungtown, p 19

Conklin paper mill, p 10

Nearing the harbor . . .

Conklin House, or Seaman's Railroad House, 75 Main, with paper mill plaque.

Bedlam Street, once Holmes Bakery, 90–94 Main.

Valentine–Denton House, 60 Main, with plaque and adjacent garden.

Main Steet shop, thought to be newly completed c 1850 for whaleman Isaac Price, and part of today's Heritage Candle Shop, 29 Main.

Burr-Gardiner House, 5-11 Main.

Post Office and telegraph office, corner Main Street and Shore Road, today Vinnie's Barbershop.

(29 Main), said to have been built c 1850 for whaleman Isaac Price, reveals clear details of a late-century remodeling. Note the wonderfully non-parallel lines of its old door, and the door handle hardware – ornamentally cast with sunflowers – that identifies its newer style.

The *barbershop* (corner Main and Shore Road), in its trim one-story building, was simultaneously a post office, telephone switchboard, and Western Union telegraph office around the turn of the century when it was new. The bracket-ended shingle pattern in the gable field is distinctive; so is the trim of incised waves around the windows and doors. Legend says this building came here, drawn by oxen from Huntington, in fair resolution of a debt.

On the north side of the street near the corner, the *Burr-Gardiner house* (5-11 Main) joins old and new. The section near the corner was built by well-known antiques dealer Valdemar Jacobsen, 1977-78, the next sections to the east in the early 1800s, and the easternmost as a separate structure c 1860. In the eastern, 1860s, section, leather worker Elbert Burr lived and housed boarders. A later occupant, Townsend Gardiner, also a shoemaker and a fisherman and laborer as well, continued the boarding-house tradition of the house. It is thought that President Lincoln's assassination was announced here to the residents of Cold Spring Harbor, on April 15, 1865, by the driver of the eastbound stage.

Spring Street and Shore Road

Right at the corner where the western end of Main Street gives way to Harbor Road (to the left) and Shore Road (down to the right), the handsome, red-brick *Cold Spring Harbor Library* building stands across from the Firehouse on land that was, in early days, the marshy edge of Cold Spring Stream. Built for the Library in 1912, the building is a crisp and restrained example of the Colonial Revival style. Today it houses the SPLIA Gallery. Behind it, the tree-studded Library Park occupies territory that was in part reclaimed by landfill.

Spring Street, not surprisingly, is named for the sweetwater springs that also name the town and marks the neighboring site of a Matinecock village.

The two-story, gable-ended *Burr-Gardiner Annex* (6 Shore Road, at Spring) once housed overflow boarders from the Main Street establishment. The building is essentially Greek Revival in style, but displays a feature of the much rarer, contemporaneous, Egyptian Revival. To see it, look at the front door. Although its three-part enframement with mitered "ears" at the top, transom,

Spring Street and Shore Road

Burr-Gardiner Annex, which served as the last CSH Customs House, 6 Shore Road.

CSH Library Building, today the SPLIA Gallery. The Library was founded c 1886 and raised $128.96 from two theater presentations, plus $75 from the sale of a donated piano, to begin its work. The Cold Spring Harbor Village Improvement Society was founded in 1899 to take the next vital steps in the Library's growth. . . . Liveryman John Totten's house, 27 Spring Street, across from Totten stables site.

and sidelights are Greek, the way the door jambs (side pieces) "batter" or widen from top to bottom, in imitation of an Egyptian gateway, denotes the more exotic style. The angular "pine tree window" in the front gable field dates to a later retrimming than the doorway. A nineteenth-century tenant of this house, Reuben Hall, was the last Collector of Customs in Cold Spring Harbor, ending the village's run, which began in 1799, as a federal Port of Delivery with its own Customs House.

On the north side of Spring Street, the remnant of an old post-and-rail fence marks the site of the *Totten stable*, the nineteenth-century village livery. Story says that whenever the stage reached the livery with its packet of letters, one bell would be sounded to summon people for their mail. On the day of Lincoln's assassination, the story goes, the mail bell tolled and tolled. The *John Totten house* (27 Spring), across the street on the south side, appears to have been built in the 1870s or 1880s. It has board-and-batten siding on the front wall, trefoil-shaped trim at its gable end, and a handsome Greek Revival front doorway.

Shore Road here leads northward from the village, along the harbor's eastern edge. The site of the 1782 Conklin Paper Mill is on the northeast (right) side of the road a short distance downhill from a big white farmhouse overlooking the harbor. It is difficult to walk along this narrow and curving road, but a slow drive yields scores of hints of the past. . . .

Sail lofts and warehouses once ranged along the shore here. Dole's Shipyard stood just before the tidal lagoon, and diarist Helen Rogers describes Commodore Vanderbilt's steam sidewheeler the *North Star* being burned for scrap right here in 1865. A little beyond, near where the oil tanks now stand, was the nineteenth-century *Eagle Dock* – a landing for coastal schooners and the sidewheel steamer *American Eagle*. Abrams Shipyard buildings stood on shore just south of the dock.

Along this stretch of Shore Road, across the road, several large, early houses command views of the harbor from the eastern hillside. One of these, the *Titus-Tavern house*, once had a sailors' bar in its above-grade basement, with steps running invitingly down to the level of the harbor and its landings.

Its near neighbor, thought also to be a Titus house, has this family name scratched in one of its old windows. A spring wells up in its basement and is carefully bricked over and piped underground to its perpetual outpouring along the harbor's edge; it is said that passersby have quenched their thirsts there for many generations. More recently, as the house's current owner tells it, local oystermen kept their catches fresh on that cool brick basement floor. To keep the oysters from "relaxing" and opening up, a turtle kept them company by crawling back and forth among them!

Further along Shore Road, near the spot where the public road ends today, the Cold Spring Harbor Beach Club centers upon on an airy, two-story building of yellow-painted shingle. This was the 1890 *Casino* – today the only architectural remnant of the exuberant *Glenada Hotel* just up the hill. The *Forest Lawn Hotel* ranged up the hillside nearby, just across the street from Eagle Dock Community Beach.

Harbor Road, the inner harbor, and Bungtown

Harbor Road borders the eastern shoreline south of the village, running down along the inner harbor to the series of ponds at its head. Harbor Road, too, has a maritime history that can still be discerned in the sites and structures bordering it.

Not far south of the village, the *Walters-Van Ausdall Hotel* (105 Harbor Road, today the Inn on the Harbor restaurant) began as a whalers' hotel. Van Ausdall is said to have provided brothel services on the upper floors of the building; he also acted as a sailors' banker, advancing and distributing living-expense money to the absent men's families. Van Ausdall segued from the age of sail into the era of steam: he was also the proprietor of one of the area's most popular excursion-era amusement spots – Columbia Grove, out on the harbor's eastern shore.

Southward along Harbor Road, handsome early- to mid-nineteenth-century houses hug each side. Most were built during the days of coastal ships and whaling, and retrimmed in the affluent resort era with bay windows, brackets, and gables. When the houses were young, the inner harbor was a bustling place, noisy and lively with blacksmith shops, sail lofts, shipyards, and stores.

About halfway between the village and the intersection at Fish Hatchery Hill, Harbor Road enters the historic core of the Jones family industries. It was here that by the early 1800s, John Hewlett and Walter Restored Jones, together with members of the Hewlett family, had established an interlocking network of local industries. Among their enterprises: woolen mills and gristmills, shipyards and repair yards, a general store, and a barrel factory.

Harbor Road,
the inner harbor, and Bungtown

Hooper House, on CSH Laboratory grounds off Bungtown Road, built in the whaling era to house workers at the nearby Jones barrel factory.

Davenport House, corner Route 25A and Bungtown Road, built for the Fish Hatchery's first superintendent in 1884.

The trim brownstone foundation of the Hewlett-Jones gristmill can still be seen here. Nearby, the handsome white clapboard Hewlett House "Harbor View," built in 1824 for Jacob C. Hewlett, still stands today overlooking the inner harbor.

The Jones enterprises rounded the head of the harbor and extended up along the west side, many of them clustering in the settlement known then and now as *Bungtown*. En route to Bungtown Road (today the entrance to the Cold Spring Harbor Laboratory) is *St. John's Episcopal Church,* built in 1836, with its tall steeple used by whaling captains as a bearings landmark for entering Cold Spring Harbor from the Sound.

Here too is the *Cold Spring Harbor Fish Hatchery & Aquarium,* founded in 1883 in a former Jones Company mill. Handsome *Davenport House,* 1884, just across Route 25A and restored by the Laboratory in 1980 in its original exuberant colors, was once the Fish Hatchery superintendent's house.

In along Bungtown Road, and well up the side of the wooded Cannon Hill, stood the signal cannon that was fired to herald the arrival of homecoming whalers. On its 107-acre grounds the Cold Spring Harbor Laboratory has meticulously restored seven buildings from the whaling era and earlier, among them: *Osterhout cottage*, an early-nineteenth-century residence; the *Wawepex building*, a whaling-era warehouse; *Hooper house*, built as workers' housing for the Jones barrel factory that stood nearby to the north; and, closer to the water, the c 1830 *Yellow House;* a Jones-era cooper shop operated nearby.

Out at the end of Bungtown Road, just below the beach historically known as "the sandspit" or "Major Jones' Beach," stands the elegant, early-Federal-style *Airslie*. This is the graceful, gambrel-roofed house built in 1806 for Major William Jones, a gentleman farmer and first cousin to the two Jones brothers who so vigorously helped the town to grow. From 1943 to the present, the directors of the Cold Spring Harbor Laboratory have made their home in Airslie.

. . .

Cold Spring village and its harbor are indelibly imprinted by the past. One era thrives in mind and heart; another leaves Victorian extravagances in scrolls and brackets and bays. Of the great resorts of summertime pleasure little remains to see, except some land filled for recreation, some tableware and silver in private and museum collections, a handsome old casino transformed to a modern beach club. Yet it is here, still, so much of the past – embedded in marshes and springs, spits of sand, and streets; in houses and stores, mills and docks. The past is rediscovered through stories told to children who tour Main Street with museum guides, through tales shared with visitors who climb to a madam's bedroom high in an old hotel, through respect paid to buildings that housed workers in a nineteenth-century cooperage. The past, in a place like Cold Spring Harbor, is in the very air.

Legacies

Founded 1883
Open every day, 10–5
· · ·

Box 535, Rte 25A
CSH 11724
516-692-6768

Cold Spring Harbor
Fish Hatchery & Aquarium

One hundred and sixteen years old this year, the Fish Hatchery was established in 1883 by the New York Fisheries Commission to help stock the lakes and streams of the state's southern region for recreational fishing. The location was ideal, thanks to the abundance of fresh water from a string of former mill ponds (created by damming the Cold Spring River generations before), and to the many sweetwater springs for which the area was famous (*see page 46*) ★ The proximity of a tidal estuary, giving access to both brackish and salt water, gave the operations added scope. The Hatchery's first Superintendent, New York's distinguished piscatologist Frederic Mather, was the first to import Brown Trout eggs into the United States from Germany; they came by ship and by rail in milk cans in 1883. ★ The Hatchery changed hands in 1982 when New York State decided to close it in favor of more modern facilities upstate. A Friends group stepped forth to keep it open – as the Cold Spring Harbor Fish Hatchery. ★ Today the Walter L. Ross II Aquarium Building, and the smaller wooden clapboard Julia Fairchild Exhibit Building, together house more than fifty different species of native freshwater fish, reptiles, and amphibians – New York State's largest collection. ★ Other facilities include a hatch house and rearing pools, young trout ponds, round ponds for Brook, Brown, and Rainbow trout, an indoor freshwater stream exhibit, and a tidal raceway where children fish for trout on special festival occasions. ★ The Hatchery today welcomes 50,000 visitors a year, including 10,000 students, exploring the freshwater ecosystems of New York State.

Cold Spring Harbor Laboratory

The Brooklyn Institute of Arts and Sciences established its new Biological Laboratory at Cold Spring Harbor in 1890 (*see page 46*), influenced by the advice of its board member Fish Commissioner Eugene Blackford, and his fellow trustee Dr Franklin Hooper, who had attended Professor Louis Agassiz's famous summer biology school near Cape Cod, Massachusetts. ★ The first lectures and researches took place in the Fish Hatchery building, until local benefactor John D. Jones contributed land and funds toward a laboratory facility in 1892. Students and lecturers were housed in homes along the harbor's western shore, once filled with Bungtown workmen of a bygone era. ★

Founded 1890
Community tours –
Saturday by appointment
· · ·

1 Bungtown Rd.
CSH 11724
516-367-8455

continued

Legacies

continued

The Carnegie Institution of Washington (CIW) established a presence here in 1904 with its Station for Experimental Evolution. Dr Charles Benedict Davenport, who had been running the summer Bio Lab, was the Station's first director. He grafted a Eugenics Record Office onto the Station, which became the Department of Genetics of the CIW and later merged with the Bio Lab. ★ The newly named Cold Spring Harbor Laboratory (of Quantitative Biology) – the institution so well known today – was formed after CIW's departure and managed by a board of prominent university researchers and interested neighbors. ★ Today, the Lab is a private, non-profit research and educational institution that focuses primarily on cancer research, neurobiology, and plant biology. It hosts scientific meetings, and conducts educational programs for the public and students at all levels, including specialized postdoctoral training for scientists. In November 1998, the Lab received state accreditation to begin an innovative PhD program – the Watson School of Biological Sciences.

Cold Spring Harbor Whaling Museum

One century after the incorporation of the Cold Spring Whaling Company, the Whaling Museum Society was founded in 1936 by two leading village residents – Dr Charles B. Davenport, recently retired from leading the Carnegie Genetics Department, and Dr Robert Cushman Murphy, Curator of Birds at New York's Museum of Natural History. ★ The Whaling Museum building was completed six years later, specifically to house the Society's prize exhibit: a nineteenth-century whaleboat – complete with original harpoons and lines – from the whaling brig *Daisy*, in which Dr Murphy himself had sailed as naturalist. ★ The Museum is one of only two accredited maritime museums in New York State, and the only facility dedicated to Cold Spring Harbor's whaling era and local history. ★ Its collections include – beyond the celebrated whaleboat-whaling implements, ships' gear and navigational instruments, a renowned scrimshaw collection, marine paintings, ship models, a meticulously accurate handmade diorama of the harbor c 1850, and a family activity room with interactive exhibits and such intriguing items as whale baleen and a sperm whale jaw. ★ Recent changing exhibits include the "Sailing Circle" (focusing on whaling wives), "Cold Spring Harbor: Time Measured by the Tide," and "CSH Whalers Around the World," as well as the popular permanent exhibits "Mark Well the Whale" and "Wonder of Whales." The Museum attracts visitors from nearly every state in the union, as well as guests from more than forty other countries, as yearly visitation reaches toward 55,000.

Founded 1936
Open 11-5 Tues-Sun;
also open Mon
Memorial Day
to Labor Day

. . .

PO Box 25, Main St.
at Turkey Lane
CSH 11724
631-367-3418

Founded 1948
Gallery Open 11-4,
seasonal schedule
516-367-6295

. . .

PO Box 148
161 Main St.
CSH 11724
631-692-4664

. . .

Historic House Museums,
seasonal hours

Society for the Preservation of Long Island Antiquities – SPLIA

The idea of a Long Island preservation society based on the model of the Society for the Preservation of New England Antiquities was the yield of meetings held at the Metropolitan Museum of Art, in 1948, by Long Islanders concerned about the destructive effects of post-war development on the region's cultural heritage. ★ For the last half century, SPLIA has been devoted to the preservation and interpretation of Long Island's past, and pursues this goal through its historic house museums (including the nearby Joseph Lloyd Manor on Lloyd Neck), an outstanding collection of Long Island decorative arts, publications on Long Island's history and architecture, educational programs, and preservation advocacy. ★ The Society owns many watercolors by Edward Lange, a Greenlawn artist who was one of the most accurate recorders of later nineteenth-century life on Long Island, and recently acquired James Bard's oil painting of the elegant steam sidewheeler *Seawanhaka*. SPLIA presents changing exhibits on Long Island's past in its Gallery at the Main Street / Shore Road corner, which for sixty-five years served as the Cold Spring Harbor Library. ★ In 1998, SPLIA headquarters moved from Setauket to the former Cold Spring Harbor Methodist Church, which it has creatively adapted for use as offices, library, and lecture hall.

Founded 1988
Open 10-4 Mon-Fri
12-4 Sat

. . .

Center at 334 Main St.

. . .

1 Bungtown Rd
CSH 11724
516-367-5170

DNA Learning Center

Now just eleven years old, the DNA Learning Center was the brainchild of David Micklos, a professional educator who first came to the CSH Laboratory as its public information officer. In 1985 he founded a DNA Literacy Program under Lab sponsorship to train high school biology instructors to teach the kind of DNA science being done at Cold Spring Harbor. The program went national a year later by means of "Vector Vans" – mobile DNA labs that crisscross the country offering hands-on summer workshops to teachers. ★ The Learning Center opened in 1988 in a Georgian-style 1920s former schoolhouse, on the site of the 1845 East Side School. Its first public presentation was "The Search for Life: Genetic Technology in the 20th Century," fresh from its first-run showing at Washington's National Museum of American History. ★ Current programs include the multi-media presentation "Long Island Discovery," created in partnership with Cablevision Systems Corporation, as well as exhibits on the human growth hormone HGH and on Nobelist Barbara McClintock. ★ The Center is best known for its hands-on approach to learning about the molecules of life, achieved through outreach programs on and off site for 14,000 students each year – and including training for high school and college faculty. In summer outreach workshops, students can progress from elementary concepts of cell biology and genetics to examining their own DNA fingerprints and sequences. Student and teacher participation topped 31,000 in 1998.

Selected Bibliography

Albion, Robert Greenhalgh. *The Rise of New York Port*, 1815-1860. New York: Charles Scribner's & Sons, 1939.

Brosky, Keriann Flanagan. *Huntington's Hidden Past*. Huntington: Maple Hill Press, 1995.

_____ . *Huntington's Past Revisited*. Huntington: Maple Hill Press, 1997.

Bunce, Mildred E. "A Cold Spring Harbor Childhood." *Long Island Forum*, Sept.–Nov. 1966.

Cold Spring Harbor Village Improvement Society. *Cold Spring Harbor Soundings*, 1953.

Earle, Walter K. *Out of the Wilderness*. Cold Spring Harbor: Whaling Museum Society, Inc., 1966.

Langhans, Rufus B. *Huntington's Historic Markers*. Huntington: Town Historian, undated.

Long Island Rail Road. "117 Years of Long Island Railroading." *Long Island Railroader*, Sept. 1952, Oct. 1952.

McKay, Richard C. *South Street – A Maritime History of New York*. New York: G. P. Putnam's Sons, 1934.

Newsday. *Long Island – Our Story*. Written/edited by staff of *Newsday*. Melville: Newsday, Inc., 1998.

Overton, Jacqueline. *Long Island's Story*. Garden City: Doubleday Doran & Co., 1929.

Peckham, Leslie E. *Clamtown*. Cold Spring Harbor, 1962.

Stephens, W. P. *Seawanhaka Corinthian Yacht Club – Origins and Early History*. New York: SCYC, 1963.

Schmitt, Frederick P. *Mark Well the Whale*. Port Washington: Ira J. Friedman Division, Kennikat Press, 1971.

Seyfried, Vincent F. *The Long Island Rail Road – A Comprehensive History*, Part 3, 1863-1880. Garden City, 1966.

Symington, Charles J. *Skippin' the Details*. New York: Clarke & Way, Inc., 1966.

Vagts, Christopher R. *Huntington at the Turn of the Century*. Huntington: Huntington Historical Society, 1974.

Valentine, Harriet G. *The Window to the Street*. Smithtown: Exposition Press, Inc., 1981.

_____ . *Main Street, Cold Spring Harbor*. Huntington: Huntington Historical Society, 1960.

Watson, Elizabeth. *Houses for Science*. Cold Spring Harbor: Cold Spring Harbor Laboratory Press, 1991.